Stefan Friedenberg

Extensions of Abelian Groups

Stefan Friedenberg

Extensions of Abelian Groups

and Torsion-Free Pairs

Südwestdeutscher Verlag für Hochschulschriften

Impressum/Imprint (nur für Deutschland/ only for Germany)
Bibliografische Information der Deutschen Nationalbibliothek: Die Deutsche Nationalbibliothek verzeichnet diese Publikation in der Deutschen Nationalbibliografie; detaillierte bibliografische Daten sind im Internet über http://dnb.d-nb.de abrufbar.

Alle in diesem Buch genannten Marken und Produktnamen unterliegen warenzeichen-, markenoder patentrechtlichem Schutz bzw. sind Warenzeichen oder eingetragene Warenzeichen der jeweiligen Inhaber. Die Wiedergabe von Marken, Produktnamen, Gebrauchsnamen, Handelsnamen, Warenbezeichnungen u.s.w. in diesem Werk berechtigt auch ohne besondere Kennzeichnung nicht zu der Annahme, dass solche Namen im Sinne der Warenzeichen- und Markenschutzgesetzgebung als frei zu betrachten wären und daher von jedermann benutzt werden dürften.

Verlag: Südwestdeutscher Verlag für Hochschulschriften Aktiengesellschaft & Co. KG
Dudweiler Landstr. 99, 66123 Saarbrücken, Deutschland
Telefon +49 681 37 20 271-1, Telefax +49 681 37 20 271-0
Email: info@svh-verlag.de
Zugl.: Duisburg-Essen, Universität, Diss., 2009

Herstellung in Deutschland:
Schaltungsdienst Lange o.H.G., Berlin
Books on Demand GmbH, Norderstedt
Reha GmbH, Saarbrücken
Amazon Distribution GmbH, Leipzig
ISBN: 978-3-8381-1788-1

Imprint (only for USA, GB)
Bibliographic information published by the Deutsche Nationalbibliothek: The Deutsche Nationalbibliothek lists this publication in the Deutsche Nationalbibliografie; detailed bibliographic data are available in the Internet at http://dnb.d-nb.de.

Any brand names and product names mentioned in this book are subject to trademark, brand or patent protection and are trademarks or registered trademarks of their respective holders. The use of brand names, product names, common names, trade names, product descriptions etc. even without a particular marking in this works is in no way to be construed to mean that such names may be regarded as unrestricted in respect of trademark and brand protection legislation and could thus be used by anyone.

Publisher: Südwestdeutscher Verlag für Hochschulschriften Aktiengesellschaft & Co. KG
Dudweiler Landstr. 99, 66123 Saarbrücken, Germany
Phone +49 681 37 20 271-1, Fax +49 681 37 20 271-0
Email: info@svh-verlag.de

Printed in the U.S.A.
Printed in the U.K. by (see last page)
ISBN: 978-3-8381-1788-1

Copyright © 2010 by the author and Südwestdeutscher Verlag für Hochschulschriften Aktiengesellschaft & Co. KG and licensors
All rights reserved. Saarbrücken 2010

Contents

Notations	2
Introduction	3
1 Preliminaries	**7**
1.1 Introduction of Ext	7
1.2 Elementary Properties of Ext	11
1.3 The rank of a group	16
1.4 Types	18
1.5 Ring Theoretical Background	20
1.6 Murley-Groups	23
1.7 First results on torsion-free Ext	27
2 The Finite Rank Case	**34**
2.1 On a Result of Goeters	34
2.2 The Divisible Outer-type Case	41
3 The General Case	**58**
3.1 Chain Conditions	58
3.2 B-coseparable Groups	64
4 An Application of our Theory: Torsion-free Pairs	**75**
4.1 Introduction	75
4.2 The Torsion-free Pairs Generated by Torsion-free Groups of Finite Rank	79
Bibliography	91

Notations

In this work, all groups are assumed to be abelian. Most of the formulations are standard as, e. g. in [Fu1].

List of symbols

$\mathbb{P}$	Set of primes
$\mathbb{N}$	The natural numbers
$\mathbb{Z}$	The integers
$\mathbb{Q}$	The rational numbers
$\aleph_0$	The first infinite cardinal
$\aleph_1$	The first uncountable cardinal
$o(a)$	The order of a group element a
$|A|$	The cardinality of the group A
A_p	The p-component of A
A_0	The nucleus of the group A
$t(A)$	The torsion part of the group A
$\mathfrak{Tf}$	The class of all torsion-free groups
$\mathfrak{Tff}$	The class of all torsion-free groups of finite rank
$\mathfrak{F}$	The class of all free groups
$\mathfrak{D}$	The class of all divisible groups
$\mathbb{Q}A$	The tensor product $\mathbb{Q} \otimes A$
$tp(R)$	The type of a rational group R
$IT(A)$	The inner type of the group A
$OT(A)$	The outer type of the group A

Introduction

Since R. Baer [B] introduced in 1933 the functor $\mathrm{Ext} = \mathrm{Ext}_{\mathbb{Z}}$ in abelian group theory, it has been considered extensively in the literature (see e.g. the books by Fuchs [Fu1] and [Fu2] and Eklof-Mekler [EkMe]). Recall that the Ext-functor is the first derived functor of the Hom-functor but $\mathrm{Ext}(A, B)$ can also be thought of as the set of all equivalence classes of short exact sequences of the form

$$0 \to B \to C \to A \to 0,$$

thus classifies all extensions of the group B by the group A. $\mathrm{Ext}(A, B)$ carries a natural group structure and one of the striking problems in abelian group theory and also in this thesis is to determine completely the structure of $\mathrm{Ext}(A, B)$ for various groups A and B. Obviously, one can generalize this problem in several ways. For example one can replace the integers $\mathbb{Z}$ by an arbitrary ring $\mathbb{R}$ to get the functor Ext_R. As for abelian groups methods from homological algebra are then available but if for instance R is not hereditary then the well-known induced Cartan-Eilenberg sequence does not end with 0 but continues with Ext_R^2 and so on. Thus the structure of the higher Ext-groups over general rings is interesting and has been considered in the literature as well (see e.g. [EkMe], [BaSa]). In this thesis we will concentrate on the case of abelian groups and in some cases on modules over subrings of the rational numbers.

In particular, the question when Ext vanishes has achieved much attention. In this context the famous Whitehead-problem asks whether every abelian group G satisfying $\mathrm{Ext}(G, \mathbb{Z}) = 0$ has to be free. For countable groups this is true by a result due to Stein in 1951 and independently to Ehrenfeucht in 1955, cf. [EkMe, XII., Prop. 1.2] but the general result had been open for many decades until Saharon Shelah, cf. [Sh1] and [Sh2], proved its independence in ZFC in 1977. On the one hand he showed that every Whitehead group is free in Gödel's constructible universe $(V = L)$. On the other hand, there exists Whitehead groups of cardinality $\aleph_1$ which are not free if we assume Martin's axiom and the negation of CH. A good reference for the theory of Whitehead groups (and more generally modules) is [EkMe, XII]. Going one step further, the more general and (as we will see) much more complicated question is:

When is $\operatorname{Ext}(A, B)$ torsion-free for abelian groups A and B? In particular, when is $\operatorname{Ext}(A, B) = 0$?

This question was also answered for extensions of $B = \mathbb{Z}$. It was shown by Griffith, Chase and Hausen that $\operatorname{Ext}(G, \mathbb{Z})$ is torsion-free if and only if G is coseparable, cf. [EkMe, IV., Thm. 2.13]. Recall that a group G is called coseparable if it is $\aleph_1$-free and every subgroup B of G with G/B finitely generated contains a direct summand H of G such that G/H is finitely generated. While any free group is coseparable there exist coseparable groups, which are not free provided CH holds, cf. [EkMe, XII, Cor. 2.11]. Otherwise, assuming the existence of large cardinals, every coseparable group is free, as was shown in [MeSh].

An easy argument shows that for torsion-free A the group $\operatorname{Ext}(A, B)$ is always divisible, hence can be characterized by cardinal invariants. What cardinals may appear is another challenging question and is almost completely known for the case $B = \mathbb{Z}$ assuming $(V = L)$ (see [EkMe]). Working under several set theoretic assumptions it was recently shown by Strüngmann and Shelah [ShStr] that all pathological situations may arise for the possible values of the torsion-free-rank and the p-rank of $\operatorname{Ext}(G, \mathbb{Z})$.

In this thesis we replace the group $\mathbb{Z}$ by an arbitrary group B and aim for a criterion for the torsion-freeness of $\operatorname{Ext}(A, B)$. It is easy to see that we may restrict our attention to extensions of torsion-free groups by torsion-free ones. In Section 3.2 we generalize the concept of coseparable groups. A B-generated group A is B-coseparable if, for any subgroup $U \subseteq A$ with A/U finitely B-presented, there exists a direct summand V of A such that $V \subseteq U$ and A/V is B-projective. We show that the equivalent characterizations of coseparable groups hold in an analog way for B-coseparable groups A if and only if B is a finitely faithful S-group with hereditary endomorphism ring.

In the finite rank case, R.B. Warfield Jr. showed in [Wa] that the p-ranks of

$\mathrm{Ext}(A, B)$ can be calculated by

$$r_p(\mathrm{Ext}(A,B)) = r_p(A) \cdot r_p(B) - r_p(\mathrm{Hom}(A,B)).$$

Since determining $r_p(\mathrm{Hom}(A,B))$ can be very difficult, the search for a better characterization arises in a natural way. H.P. Goeters [Goe] answered this question if the outer type of the group B is not the type of the rational numbers: In this case, $\mathrm{Ext}(A,B)$ is torsion-free if and only if the B-radical of A is p-divisible for all $p \in \mathrm{supp}(B)$ and the outer type of the group $A/K_B(A)$ is less than or equal to the inner type of B. Here $\mathrm{supp}(B)$ means all primes that B is not divisible by. In Theorem 2.1.4 we improve Goeters' result by reducing it to the following nice equivalence: In the case considered by Goeters, $\mathrm{Ext}(A,B)$ is torsion-free if and only if $OT((A \otimes B_0)/D) \leq IT(B)$, where D denotes the maximal divisible subgroup of $A \otimes B_0$ and B_0 is the nucleus of B.

To get a more general result we have to drop the condition on the outer type of B. It will turn out that this complicates the problem heavily already in very simple cases. Therefore we consider Murley groups. Such a group B has the property that $r_p(B) \leq 1$ and hence a Murley group of rank 2 always is a group with rational outer type. Even in this apparently simple case it turns out, that there is no generalization of our theorem, which can be seen in Section 2.2, especially Lemmas 2.2.17 and 2.2.19 as well as Theorems 2.2.15 and 2.2.18: Here we start with the short exact sequence

$$0 \to \mathbb{Q}_p \oplus \mathbb{Q}_p \to A \to U \to 0$$

provided $\mathrm{Ext}(U,B)$ is torsion-free. Then $\mathrm{Ext}(A,B)$ is torsion-free if and only if $K_B(A) = 0$ and $r_0(\mathrm{Hom}(A,B)) > r_0(\mathrm{Hom}(U,B)) + 1$; and $r_0(\mathrm{Hom}(A,B)) > r_0(\mathrm{Hom}(U,B)) + 2$, respectively, depending on the structure of the Murley group B.

In the infinite rank case one might expect similar results for torsion-free extensions as in the case of vanishing Ext-groups. The search for a Γ-invariant seems natural but turns out to be not promising at all. Let $A = \bigcup_{\alpha < \kappa} A_\alpha$. We show that $\mathrm{Ext}(A,B)$ is a torsion-free group if and only if $\mathrm{Ext}(A_{\alpha+1}/A_\alpha, B)$ is torsion-free for all $\alpha < \kappa$ and the filtration is B-cobalanced provided the existence of supercompact cardinals (see Thm. 3.1.4 to Thm. 3.1.8). But in this model of set theory A turns out to be free if $B = \mathbb{Z}$ but we do not know what happens for general B.

We end with an application of our results. We generalize the concept of cotorsion pairs, introduced by L. Salce in 1977 [Sa], to torsion-free pairs. While cotorsion pairs deal with the case $\mathrm{Ext}(A,B) = 0$, our torsion-free pairs consider groups A such that $\mathrm{Ext}(A,B)$ is torsion-free for all groups B in some classes of groups $\mathcal{B}$. Here we turn our attention to torsion-free pairs singly generated by rational groups, which build a complete lattice anti-isomorphic to the lattice of types (see Thm. 4.2.8).

Chapter 1

Preliminaries

1.1 Introduction of Ext

There are several ways to introduce the functor Ext in the category of abelian groups. We choose a way for our considerations that needs only a minimum of knowledge. We describe Ext with the help of short exact sequences.

Definition 1.1.1 *A sequence of groups A_i and homomorphisms α_i*

$$A_0 \xrightarrow{\alpha_1} A_1 \xrightarrow{\alpha_2} A_2 \cdots \xrightarrow{\alpha_k} A_k$$

with $k \geq 2$ is called **exact at A_i** *if $\operatorname{Im}(\alpha_i) = \operatorname{Ker}(\alpha_{i+1})$. It is called* **exact** *if it is exact at any place. An exact sequence of the form*

$$0 \to A \to B \to C \to 0$$

is called a **short exact sequence**.

There are some elementary properties of short exact sequences:

Theorem 1.1.2 *Let*

$$0 \to A \xrightarrow{\alpha} B \xrightarrow{\beta} C \to 0 \qquad (E)$$

be a short exact sequence. Then the following hold:

i) *α ist injektiv;*

ii) *β ist surjektiv;*

iii) *$C \cong B/\operatorname{Im}(\alpha)$.*

Proof: See [Fu1, Par. 2]. □

The technical lemma below is frequently used in homological algebra.

Lemma 1.1.3 (5-Lemma) *The following hold for an exact diagram*

$$
\begin{array}{ccccccccc}
A_1 & \xrightarrow{\alpha_1} & A_2 & \xrightarrow{\alpha_2} & A_3 & \xrightarrow{\alpha_3} & A_4 & \xrightarrow{\alpha_4} & A_5 \\
\downarrow{\gamma_1} & \circlearrowleft & \downarrow{\gamma_2} & \circlearrowleft & \downarrow{\gamma_3} & \circlearrowleft & \downarrow{\gamma_4} & \circlearrowleft & \downarrow{\gamma_5} \\
B_1 & \xrightarrow{\beta_1} & B_2 & \xrightarrow{\beta_2} & B_3 & \xrightarrow{\beta_3} & B_4 & \xrightarrow{\beta_4} & B_5
\end{array}
$$

i) *If γ_1 is surjective and γ_2, γ_4 are injective then also γ_3 is injective.*

ii) *If γ_5 is injective and γ_2, γ_4 are surjective then γ_3 is surjektiv, too.*

iii) *If γ_1 is surjective, γ_5 is injective and γ_2, γ_4 are bijective then also γ_3 is bijective.*

Proof: See [Fu1, Lemma 2.3]. □

We also need the definition of a splitting short exact sequence.

Definition 1.1.4 *A short exact sequence*

$$0 \to A \xrightarrow{\alpha} B \xrightarrow{\beta} C \to 0$$

splits *if $\mathrm{Im}(\alpha)$ is a direct summand of B.*

Splitting sequences can be characterized by

Theorem 1.1.5 *The following are equivalent for a short exact sequence*

$$0 \to A \xrightarrow{\alpha} B \xrightarrow{\beta} C \to 0 \qquad (E)$$

i) *(E) splits.*

ii) *There is $\gamma \in \mathrm{Hom}(B, A)$ with $\gamma \circ \alpha = id_A$.*

iii) *There exists $\delta \in \mathrm{Hom}(C, B)$ with $\beta \circ \delta = id_C$.*

Proof: See [Fu1, Par. 50]. □

Now we can define an equivalence-relation on the short exact sequences beginning with A and ending with C.

Definition 1.1.6 *The short exact sequences*

$$0 \to A \to B \to C \to 0 \qquad (E)$$

and

$$0 \to A \to B' \to C \to 0 \qquad (E')$$

are called **equivalent**, *denoted by* $(E) \equiv (E')$ *if there is a homomorphism* $\beta : B \to B'$, *such that the diagram*

$$\begin{array}{ccccccccc}
0 & \to & A & \to & B & \to & C & \to & 0 \\
& & \| id_A & \circlearrowleft & \downarrow \beta & \circlearrowleft & \| id_C & & \\
0 & \to & A & \to & B' & \to & C & \to & 0
\end{array}$$

commutes.

It is easy to see, that the homomorphism β is already an isomorphism by the 5-Lemma. Hence the introduced term of equivalence is really an equivalence-relation.

It takes some work to define an additive group structure on the equivalence-classes of short exact sequences. This group is denoted by $\mathbf{Ext}(\mathbf{C}, \mathbf{A})$. $\mathrm{Ext}(C, A)$ is called the **extension group of A by C**. To get a rough idea of the concept of the composition in this group one should have a look at the following constructions: If we consider $\gamma \in \mathrm{Hom}(C', C)$ and the given short exact sequence

$$0 \to A \to B \to C \to 0 \qquad (E)$$

then we can construct a short exact sequence $(E\gamma)$ as a pullback of (E) by

$$\begin{array}{ccccccccc}
0 & \to & A & \to & B' & \to & C' & \to & 0 \qquad (E\gamma) \\
& & \| id_A & \circlearrowleft & \downarrow \beta & \circlearrowleft & \downarrow \gamma & & \\
0 & \to & A & \to & B & \to & C & \to & 0 \qquad (E)
\end{array}$$

In a similar way we obtain the short exact sequence (αE) as a pushout for $\alpha \in \mathrm{Hom}(A, A')$ by

$$\begin{array}{ccccccccc}
0 & \to & A & \to & B & \to & C & \to & 0 \qquad (E) \\
& & \downarrow \alpha & \circlearrowleft & \downarrow \beta & \circlearrowleft & \| id_C & & \\
0 & \to & A' & \to & B' & \to & C & \to & 0 \qquad (\alpha E)
\end{array}$$

Now we define the direct sum of two short exact sequences (E_i)

$$0 \to A_i \xrightarrow{\mu_i} B_i \xrightarrow{\nu_i} C_i \to 0 \qquad (E_i)$$

with $i = 1, 2$ by
$$0 \to A_1 \oplus A_2 \xrightarrow{\mu_1 \oplus \mu_2} B_1 \oplus B_2 \xrightarrow{\nu_1 \oplus \nu_2} C_1 \oplus C_2 \to 0 \qquad (E_1 \oplus E_2).$$

Furthermore we specify the homomorphisms γ and α by $\Delta_C : C \to C \oplus C$ via $c \mapsto (c, c)$ and $\nabla_A : A \oplus A \to A$ via $(a_1, a_2) \mapsto a_1 + a_2$. Hence we can define an addition in the group $\mathrm{Ext}(C, A)$ by $[E_1] + [E_2] = [E_1 + E_2]$. Here we have
$$E_1 + E_2 = \nabla_A (E_1 \oplus E_2) \Delta_C.$$

More details on the group structure of extension groups can be found in [Fu1, Par. 50].

The following proposition is essential and easily deduced from Theorem 1.1.5.

Proposition 1.1.7 *Any short exact sequence*
$$0 \to A \to B \to C \to 0$$
splits if and only if $\mathrm{Ext}(C, A) = 0$.

Proof: See [Fu1, Par. 50]. $\square$

We end this section with a technical result, which frequently will be used.

Lemma 1.1.8 *Let* $(E) \quad 0 \to A \xrightarrow{\alpha} B \xrightarrow{\beta} C \to 0$ *be a short exact sequence such that the induced sequence*
$$0 \to \mathrm{Hom}(A, A) \xrightarrow{\alpha'} \mathrm{Hom}(B, A) \xrightarrow{\beta'} \mathrm{Hom}(C, A) \to 0$$
is a splitting exact sequence. Then also (E) *splits.*

Proof: Since the induced sequence splits also the top row of the commutative diagram
$$\begin{array}{ccccccccc}
0 & \to & \mathrm{Hom}(H_A(A), A) & \xrightarrow{\alpha^*} & \mathrm{Hom}(H_B(A), A) & \to & \mathrm{Hom}(H_C(A), A) & & \\
& & \uparrow \psi_A & & \uparrow \psi_B & & \uparrow \psi_C & & \\
0 & \to & A & \xrightarrow{\alpha} & B & \to & C & \to & 0,
\end{array}$$
does, in which ψ_A is an isomorphism.
Now let $\sigma : \mathrm{Hom}(H_B(A), A) \to \mathrm{Hom}(H_A(A), A)$ with $\sigma \circ \alpha^* = id_{\mathrm{Hom}(H_A(A), A)}$. Since
$$\psi_A^{-1} \circ \sigma \circ \psi_B \circ \alpha = \psi_A^{-1} \circ \sigma \circ \alpha^* \circ \psi_A = \psi_A^{-1} \circ \psi_A = id_A,$$
the bottom-row splits. $\square$

1.2 Elementary Properties of Ext

A very important tool in our investigations of extension groups is

Lemma 1.2.1 (Cartan-Eilenberg) *Let*

$$0 \to A \xrightarrow{\alpha} B \xrightarrow{\beta} C \to 0 \qquad (E)$$

be a short exact sequence and H an arbitrary group. Then also the sequences

$$0 \to \operatorname{Hom}(H,A) \to \operatorname{Hom}(H,B) \to \operatorname{Hom}(H,C)$$
$$\xrightarrow{E_*} \operatorname{Ext}(H,A) \xrightarrow{\alpha_*} \operatorname{Ext}(H,B) \xrightarrow{\beta_*} \operatorname{Ext}(H,C) \to 0$$

and

$$0 \to \operatorname{Hom}(C,H) \to \operatorname{Hom}(B,H) \to \operatorname{Hom}(A,H)$$
$$\xrightarrow{E^*} \operatorname{Ext}(C,H) \xrightarrow{\beta^*} \operatorname{Ext}(B,H) \xrightarrow{\alpha^*} \operatorname{Ext}(A,H) \to 0$$

are exact.

The definition of the connecting maps E_* and E^* can be found in [Fu1, Par. 51], as well as the other homomorphisms.

Proof: See [Fu1, Thm. 51.3]. □

We now consider closure properties of the functor Ext and determine the projectives and injectives. We also procure some more informations about extension groups.

Lemma 1.2.2 *If $\operatorname{Ext}(C,A) = 0$, $C' \leq C$ and A' is an epimorphic image of A then also $\operatorname{Ext}(C',A') = 0$.*

Proof: Clear using Lemma 1.2.1. □

Lemma 1.2.3 *The following hold for any group A:*

i) *If F is a free group then $\operatorname{Ext}(F,A) = 0$.*

ii) *If H is a divisible group then $\operatorname{Ext}(A,H) = 0$.*

Proof: See [Fu1, Par. 52 (A) and (B)]. □

The next lemma shows, that Ext commutes with direct sums in the first component and direct products in the second component.

Lemma 1.2.4 *There are natural isomorphisms*

i) $\operatorname{Hom}(\bigoplus_{i \in I} C_i, A) \cong \prod_{i \in I} \operatorname{Hom}(C_i, A)$

ii) $\operatorname{Hom}(C, \prod_{j \in J} A_j) \cong \prod_{j \in J} \operatorname{Hom}(C, A_j)$

and

iii) $\operatorname{Ext}(\bigoplus_{i \in I} C_i, A) \cong \prod_{i \in I} \operatorname{Ext}(C_i, A)$

iv) $\operatorname{Ext}(C, \prod_{j \in J} A_j) \cong \prod_{j \in J} \operatorname{Ext}(C, A_j)$

for arbitrary index sets I and J.

Proof: See [Fu1, Thm. 43.1 and 43.2] and [Fu1, Thm. 52.2]. □

In some cases the computation of Ext is very simple. Therefore recall

Definition 1.2.5 *For any group A and any $n \in \mathbb{N}$ we call*

$$A[n] := \{a \in A \mid na = 0\}$$

*the **n-socle of A**.*

We can now easily see

Lemma 1.2.6 *For any group A and $n \in \mathbb{N}$ the following hold:*

i) $\operatorname{Ext}(\mathbb{Z}/n\mathbb{Z}, A) \cong A/nA$.

i) $\operatorname{Ext}(A, \mathbb{Z}/n\mathbb{Z}) \cong \operatorname{Ext}(A[n], \mathbb{Z}/n\mathbb{Z})$.

Proof: See [Fu1, Par. 52 (D) and (F)]. □

Another important result is given by the next lemma. This is the basic motivation of this thesis. Recall that divisible groups A are of the form

$$A = \bigoplus_{r_0} \mathbb{Q} \oplus \bigoplus_{p \in \mathbb{P}} [\bigoplus_{r_p} \mathbb{Z}_{p^\infty}]$$

for some cardinals r_0 und r_p.

Lemma 1.2.7 *If A is torsion-free then $\operatorname{Ext}(A, B)$ is divisible for any group B.*

Proof: Although the proof can be found in [Fu1, Par. 52 (I)], we represent it here for the convenience of the reader.
Since A is torsion-free we have $A[m] = 0$ for any $m \in \mathbb{N}$, which implies that $0 \to A \xrightarrow{m} A$ is exact. Hence also the sequence $\operatorname{Ext}(A,B) \xrightarrow{m} \operatorname{Ext}(A,B) \to 0$ is exact, which implies that $\operatorname{Ext}(A,B)$ is m-divisible. $\square$

Moreover, it will be usefull to know

Lemma 1.2.8 *Let A be a p-group and B be p-divisible. Then* $\operatorname{Ext}(A,B) = 0$.

Proof: See [Fu1, Par. 52 (K)]. $\square$

Our main interest is to investigate when $\operatorname{Ext}(A,B)$ is torsion-free for torsion-free groups A and B. We would like to give some reason, why we restrict ourselves to torsion-free groups. Therefore we distinguish several possible cases.

But afore we need the concept of pure subgroups.

Definition 1.2.9 *A subgroup B of A is called **pure** if*

$$nB = B \cap nA$$

*for all $n \in \mathbb{Z}$. Pure subgroups are denoted by $\mathbf{B} \leq_* \mathbf{A}$. For a prime $p \in \mathbb{P}$ the subgroup B is called **p-pure** if*

$$p^k B = B \cap p^k A$$

for all $k \in \mathbb{N}$.
Furthermore, if A is a torsion-free group we can define the purification B_* of B in A as the intersection of all pure subgroups of A containing B, even in the case that B is just a subset of A.

This definition is obviously equivalent to the assertion that any equality of the form $nx = a \in A$ which is solvable in A also has a solution in B. For example, direct summands are always pure subgroups. Moreover, if A/B is torsion-free then $B \leq_* A$ and if both groups are torsion-free then the converse holds.

Firstly we assume A and B to be torsion groups. Then we decompose A and B into their p-components $A = \bigoplus_{p \in \mathbb{P}} A_p$ and $B = \bigoplus_{p \in \mathbb{P}} B_p$. Therefore we have

$$\operatorname{Ext}(A,B) \cong \prod_{p \in \mathbb{P}} \operatorname{Ext}(A_p, \bigoplus_{p \in \mathbb{P}} B_p) \cong \prod_{p \in \mathbb{P}} \left[\operatorname{Ext}(A_p, B_p) \oplus \operatorname{Ext}(A_p, \bigoplus_{q \in \mathbb{P} \setminus \{p\}} B_q) \right]$$

by Lemma 1.2.4. Since $\bigoplus_{q \in \mathbb{P} \setminus \{p\}} B_q$ is p-divisible and A_p is a p-group we have

$$\operatorname{Ext}(A_p, \bigoplus_{q \in \mathbb{P} \setminus \{p\}} B_q) = 0$$

by Lemma 1.2.8 and therefore

$$\operatorname{Ext}(A,B) \cong \prod_{p \in \mathbb{P}} \operatorname{Ext}(A_p, B_p).$$

So we may restrict our attention to the case that A and B are p-groups. Since any divisible group is a direct summand in every group containing it we also may assume that B is a reduced p-group by Lemmas 1.2.3 and 1.2.4. We decompose A into its divisible part D and its reduced part R. Thus $A = D \oplus R$ and hence $\operatorname{Ext}(A, B) \cong \operatorname{Ext}(D, B) \oplus \operatorname{Ext}(R, B)$. So $\operatorname{Ext}(A, B)$ is torsion-free if and only if the groups $\operatorname{Ext}(D, B)$ and $\operatorname{Ext}(R, B)$ are torsion-free.

i) When is $\operatorname{Ext}(R, B)$ a torsion-free group?

Let R' be a basic subgroup of R. Then either $R = R' = 0$ or there is $n \in \mathbb{N}$ such that $\mathbb{Z}/p^n\mathbb{Z}$ is a direct summand of R' because R is reduced. So $\mathbb{Z}/p^n\mathbb{Z}$ is a pure subgroup of R' and hence of R since basic subgroups are pure. By [Fu1, 27.5] we see that $\mathbb{Z}/p^n\mathbb{Z}$ must be a direct summand of R as it is bounded. Thus the group $\operatorname{Ext}(\mathbb{Z}/p^n\mathbb{Z}, B) \cong B/p^n B$ is isomorphic to a direct summand of $\operatorname{Ext}(R, B)$. But $B/p^n B$ is torsion and not 0 since B is reduced. So $\operatorname{Ext}(R, B)$ cannot be torsion-free except 0 and this is only the case if $R = 0$ thus A is divisible.

ii) When is $\operatorname{Ext}(D, B)$ a torsion-free group?

Here let B' be a basic subgroup of B. With similar arguments as above we see that there is some $n \in \mathbb{N}$ such that $\mathbb{Z}/p^n\mathbb{Z}$ is a direct summand of B and thus $\operatorname{Ext}(D, \mathbb{Z}/p^n\mathbb{Z}) \cong \operatorname{Ext}(D[p], \mathbb{Z}/p^n\mathbb{Z})$ is isomorphic to a direct summand of $\operatorname{Ext}(D, B)$. But $\operatorname{Ext}(D[p], \mathbb{Z}/p^n\mathbb{Z})$ contains torsion and is not 0. Hence $\operatorname{Ext}(D, B)$ cannot be torsion-free except 0.

We summarize the above in

Corollary 1.2.10 *Let A and B be torsion. Then the following hold:*

i) $\mathrm{Ext}(A, B)$ *is torsion-free if and only if* $\mathrm{Ext}(A, B) = 0$;

ii) *If A and B are p-groups then* $\mathrm{Ext}(A, B) = 0$ *if and only if $A = 0$ or $B = 0$ or B is divisible.*

Proof: Clear by the statements above. $\square$

Now let A be a mixed group and B torsion-free. We use the short exact sequence

$$0 \to t(A) \to A \to A/t(A) \to 0,$$

where $t(A)$ denotes the torsion part of A, and apply Lemma 1.2.1.
By [Fu1, Thm. 52.3] we have $\mathrm{Ext}(t(A), B) \cong \mathrm{Hom}(t(A), D/B)$, where D denotes the divisible hull of B, and hence

$$\mathrm{Ext}(A, B) \cong \mathrm{Ext}(t(A), B) \oplus \mathrm{Ext}(A/t(A), B)$$

$$\cong \mathrm{Hom}(t(A), D/B) \oplus \mathrm{Ext}(A/t(A), B).$$

Thus it is necessary that $\mathrm{Ext}(A/t(A), B)$ is torsion-free for $\mathrm{Ext}(A, B)$ to be torsion-free. However, we also consider the direct summand $\mathrm{Hom}(t(A), D/B)$. This is a reduced algebraically compact group by [Fu1, Thm. 46.1]. Therefore we may restrict our attention to the case that $t(A)$ is a p-group by similar arguments as in the case of torsion groups above. Since D/B is divisible it must be of the form $D/B \cong \bigoplus_{r_0} \mathbb{Q} \oplus \bigoplus_{p \in \mathbb{P}} \left[\bigoplus_{r_p} \mathbb{Z}_{p^\infty} \right]$ and hence we have

$$\mathrm{Hom}(t(A), D/B) \cong \mathrm{Hom}(t(A), \bigoplus_{r_0} \mathbb{Q}) \oplus \mathrm{Hom}(t(A), \bigoplus_{p \in \mathbb{P}} \left[\bigoplus_{r_p} \mathbb{Z}_{p^\infty} \right]).$$

But the first direct summand is 0 since $t(A)$ is torsion. Now let A' be a basic subgroup of $t(A)$. If $t(A)$ is divisible then $\mathrm{Hom}(t(A), D/B)$ is torsion-free. Thus we may assume that $t(A)$ is reduced. Therefore we have again $t(A) = 0$, which is trivial, or there exists $n \in \mathbb{N}$ such that $\mathbb{Z}/p^n\mathbb{Z}$ is a direct summand of $t(A)$. We distinguish two cases:

i) $r_p \neq 0$

Then the group $\mathrm{Hom}(\mathbb{Z}/p^n\mathbb{Z}, \mathbb{Z}_{p^\infty})$ is a direct summand of the group $\mathrm{Hom}(t(A), D/B)$. But this summand is not 0 and not torsion-free. Thus $\mathrm{Hom}(t(A), D/B)$ cannot be torsion-free.

ii) $r_p = 0$

Then, trivially, we have $\text{Hom}(t(A), D/B) = 0$ since $t(A)$ is assumed to be a p-group.

Alltogether we conclude that $\text{Hom}(t(A), D/B)$ is torsion-free if and only if it vanishes or $t(A)$ is divisible.

Corollary 1.2.11 *Let A be a mixed group and B torsion-free. Then $\text{Ext}(A, B)$ is torsion-free if and only if $\text{Ext}(A/t(A), B)$ is torsion-free and $t(A)$ is divisible.*

Proof: Clear. □

Next A is assumed to be torsion and B to be a mixed group. Then we have $\text{Hom}(A, B/t(B)) = 0$ and hence we get the short exact sequence

$$0 \to \text{Ext}(A, t(B)) \to \text{Ext}(A, B) \to \text{Ext}(A, B/t(B)) \to 0$$

by Lemma 1.2.1. Thus if $\text{Ext}(A, B)$ is torsion-free the group $\text{Ext}(A, t(B))$ is torsion-free as well. But A and $t(B)$ are both torsion, hence $\text{Ext}(A, t(B)) = 0$ by the statements above. In this case we conclude that also $\text{Ext}(A, B/t(B))$ is torsion-free. So it must vanish or A is divisible.

The question, if $\text{Ext}(A, B)$ has torsion elements for torsion-free A and torsion B was answered by A. Mader.

Lemma 1.2.12 *Let p be a fixed prime. If A is countable or A/pA is finite then $\text{Ext}(A, B)$ is torsion-free for any reduced p-group B.*

Proof: See [Ma1]. □

This means, that we may restrict our considerations to torsion-free groups A and B in order to determine when $\text{Ext}(A, B)$ is torsion-free.

Thus in the remainder of this thesis, all groups are assumed to be torsion-free unless stated otherwise.

1.3 The rank of a group

Our main purpose is to consider torsion-free abelian groups of finite rank. Therefore we define

Definition 1.3.1 *A system $\{a_1, a_2, \ldots, a_k\}$ of elements of a group A with $a_i \neq 0$ for all $i \in \{1, 2, \ldots, k\}$ is called* **linearly independent** *if*

$$n_1 a_1 = n_2 a_2 = \cdots = n_k a_k = 0$$

whenever

$$n_1 a_1 + n_2 a_2 + \cdots + n_k a_k = 0 \quad \text{for } n_i \in \mathbb{Z}.$$

This means $n_i = 0$ if $o(a_i) = \infty$ or $o(a_i) | n_i$ in case of finite $o(a_i)$.

The cardinalty of a maximal linearly independent system in which all elements have infinite or prime power order is called the **rank** of a group A, denoted by $\mathbf{rk(A)}$. If we restrict on elements of infinite order we call this cardinality the **torsion-free rank**, denoted by $\mathbf{r_0(A)}$. Similarly we define the **p-rank** of A, denoted by $\mathbf{r'_p(A)}$.

Proposition 1.3.2 *For any group A we have $rk(A) = r_0(A) + \sum_p r'_p(A)$.*

Proof: Follows immediately from the definitions. □

Furthermore, we also can define the p-rank for torsion-free groups:

Definition 1.3.3 *Let A be a torsion-free group. Then we call $dim_{\mathbb{Z}/p\mathbb{Z}}(A/pA)$ the* **p-rank of A** *and denote it by $\mathbf{r_p(A)}$.*

We finish this section with a well-known and frequently used result.

Lemma 1.3.4 *Let B be a subgroup of A. Then the following hold:*

i) $rk(B) \leq rk(A)$;

ii) $rk(A) \leq rk(B) + rk(A/B)$;

iii) $r_0(A) = r_0(B) + r_0(A/B)$.

Proof: See [Fu1, Par. 16]. □

1.4 Types

The subgroups of $\mathbb{Q}$, which are also called rational groups, and their direct sums play an important role in the investigation of torsion-free abelian groups. For example the localization of the integers $\mathbb{Z}$ at the prime p is given by

$$\mathbb{Q}_p := \left\{ \frac{a}{b} \in \mathbb{Q} \mid ggT(b,p) = 1 \right\} \leq \mathbb{Q}.$$

Passing to isomorphic copies we may assume, without loss of generality, $1 \in R$, or, equivalently, $\mathbb{Z} \leq R \leq \mathbb{Q}$ for any rational group R.

Definition 1.4.1 *A group A is called* **completely decomposable** *if A is of the form $A = \bigoplus_{i \in I} R_i$ where the R_i's are rational groups.*

Note that if A is a rational group we have a system of generators

$$A = \left\langle \frac{1}{p^k} \mid p \in \mathbb{P} \text{ and } k \text{ maximal with } \frac{1}{p^k} \in A \right\rangle.$$

This motivates the following definition for elements of arbitrary torsion-free groups.

Definition 1.4.2 *Let $0 \neq a \in A$ be an arbitrary element of the group A. Furthermore let $n_p^A(a) := \sup \{ n \in \mathbb{N} \text{ with } a \in p^n A \}$. Then the sequence $(n_p(a))_{p \in \mathbb{P}}$ is called the* **characteristic** *of a in A, denoted by*

$$\chi^A(a) = (n_2, n_3, n_5, \dots) = (n_p(a))_{p \in \mathbb{P}}.$$

If it is clear from the context, in which group we work, we put $n_p^A(a) = n_p(a)$.

We now define an equivalence-relation as follows: two characteristics of elements $a \in A$ are said to be **equivalent** if and only if they only differ on finitely many finite entries. It is obvious that two characteristics $\chi^A(a)$ und $\chi^A(a')$ belong to the same equivalence class provided the elements a und a' are linearly dependent. Since in the case of a rational group A any two elements are linearly dependent, there is exactly one equivalence class; this class can be denoted by

$$[(n_2, n_3, n_5, \dots)],$$

where $(n_2, n_3, n_5, \dots)$ is the characteristic of an arbitrary element of A, for example, of $1 \in A$.

Definition 1.4.3 *For any rational group A we call* $[(n_2, n_3, n_5, \ldots)]$ *the* **type** *of A which will be denoted by* **tp(A)**.

R. Baer showed that all rational groups can be classified by their types. Also note that all rank-1 groups can be identified, up to isomorphism, with a subgroup of $\mathbb{Q}$ containing $\mathbb{Z}$.

Now we can define

Definition 1.4.4 i) *A torsion-free group A is called* **strongly homogeneous** *if, for any two pure rank-1 subgroups X and Y of A, there is an automorphism of A sending X onto Y.*

ii) *We call A* **homogeneous** *if all pure rank-1 subgroups of A have the same type.*

Note that strongly homogeneous groups are homogeneous.

There is a natural generalisation of the concept of types for groups of finite rank:

Definition 1.4.5 *If A is a torsion-free group of finite rank let* $S = \{x_1, x_2, \ldots, x_n\}$ *be a maximal linearly independent subset of A. Putting*

$$X_i = \langle x_i \rangle_* \text{ and } Y_i = \langle x_1, \ldots, x_{i-1}, x_{i+1}, \ldots, x_n \rangle_*$$

we define

$$IT(A) = \inf\{tp(X_1), \ldots, tp(X_n)\}$$

and

$$OT(A) = \sup\{tp(A/Y_1), \ldots, tp(A/Y_n)\}.$$

$IT(A)$ *is called the* **inner type of** *A. Analogously, we call* $OT(A)$ *the* **outer type of** *A.*

Note, that this definition does not depend on the choice of the elements x_i.

As a simple example we consider the Corner-group of rank 2. This group A has the following properties: Any pure rank-1 subgroup B of A is free and $A/B \cong \mathbb{Q}$. So, we directly conclude that $IT(A) = tp(\mathbb{Z})$ and $OT(A) = tp(\mathbb{Q})$.

Some basic results about inner and outer types are summarized in the next lemma.

Lemma 1.4.6 *Let A and B be torsion-free groups of finite rank. Then the following hold:*

i) $r_0(\mathrm{Hom}(A,B)) \leq r_0(A)r_0(B)$. *Moreover,* $r_0(\mathrm{Hom}(A,B)) = r_0(A)r_0(B)$ *if and only if* $OT(A) \leq IT(B)$;

ii) $OT(A) = tp(\mathbb{Q})$ *if and only if* $r_p(A) < r_0(A)$ *for any* $p \in \mathbb{P}$.

Proof: See [Ar, Thm. 1.11]. □

We end this section with the definition of the nucleus of a torsion-free group:

Definition 1.4.7 *Let A be a torsion-free group. Then we call*
$$A_0 := \left\langle \frac{1}{p^n} \mid n \in \mathbb{N} \text{ and } p \in \mathbb{P} \text{ with } pA = A \right\rangle \leq \mathbb{Q}$$
*the **nucleus** of A, denoted by A_0.*
In other words, A_0 is the largest subring of $\mathbb{Q}$ such that A is still an A_0-module.

Note that $tp(A_0) \leq IT(A)$ for any torsion-free group A.

1.5 Ring Theoretical Background

In this section we will shortly summarize some well-known facts in ring theory.

Definition 1.5.1 *A torsion-free group A of finite rank is called **finitely faithful** if $IA \neq A$ for all maximal right ideals I of finite index in $\mathrm{End}(A)$. A is called an **S-group** if $\mathrm{Hom}(A,B)(A) = B$ for all subgroups B of finite index in A.*

Associated with every abelian group B is a pair (H_B, T_B) of functors between the category of abelian groups and the category M_E of right E-modules where $E = \mathrm{End}(B)$ is the endomorphism ring of B. These functors are defined by $H_B(G) = \mathrm{Hom}(B,G)$ and $T_B(M) = M \otimes_E B$ for all abelian groups G and all $M \in M_E$. There are induced natural maps $\theta_G : T_B H_B(G) \to G$ and $\phi_M : M \to H_B T_B(M)$ defined by $\theta_G(\alpha \otimes a) = \alpha(a)$ and $[\phi_M(x)](a) = x \otimes a$. The group G is **(finitely) B-generated** if and only if it is an epimorphic image of $\bigoplus_I B$ for some (finite) index set I. It is easy to see that G is B-generated if $S_B(G) = G$ where $S_B(G) := \mathrm{Im}(\theta_G)$.

With these notations we may now state another important result in the context of splitting exact sequences, which will be frequently used. The following is a generalization of Baer's Lemma:

Lemma 1.5.2 *Let A be a torsion-free group of finite rank. Then the following are equivalent:*

i) *If B is a subgroup of a group G with $G/B \cong A$ and $S_A(G) + B = G$ then B is a direct summand of B;*

ii) *If I is a maximal right ideal of $\text{End}(A)$ then $IA \neq A$.*

Proof: See [Ar, Thm. 5.6]. □

Before we continue with further results some more notations are needed. G is said to be B-**solvable** if θ_G is an isomorphism. A group P is **(finitely) B-projective** if it is a direct summand of $\bigoplus_I B$ for some (finite) index set I. If B is a torsion-free group of finite rank, then all B-projective groups are B-solvable. For a submodule U of $H_A(G)$, put $UB = \langle \phi(B) | \phi \in U \rangle$.

Let $N(E)$ denote the nilradical of E, i. e. $N(E)$ contains all elements $x \in E$ such that there is an $r \in \mathbb{Z}$ with $x^r = 0$. If B is a torsion-free group of finite rank then $N(E) = 0$ if and only if the quasi-endomorphism ring $\mathbb{Q}E$ of B is semi-simple Artinian. Observe that for such B, a right E-module M is non-singular (singular) if and only if its additive group is torsion-free (torsion). Moreover, the S-closure of a submodule of a non-singular module coincides with its $\mathbb{Z}$-purification by [Al5].

Proposition 1.5.3 *Let B be a torsion-free non-free group of finite rank with $N(E) = 0$. Then there exists an exact sequence $0 \to U \to B \to B/U \to 0$ with respect to which B is not projective.*

Proof: Let U be a full free subgroup of B. Since B is not free and $N(E) = 0$, we have $\text{Hom}(B, \mathbb{Z}) = 0$. Assume, for contradiction, that B is projective with respect to
$$0 \to U \to B \to B/U \to 0.$$
Then the induced sequence
$$0 = H_B(U) \to H_B(B) \to H_B(B/U) \to 0$$
is exact. In particular, $H_B(B/U)$ is a countable torsion-free group.
On the other hand, we have the exact sequence
$$0 \to \text{Hom}(B/U, B/U) \to \text{Hom}(B, B/U) = H_B(B/U).$$

Since B/U is torsion, $\text{Hom}(B/U, B/U)$ cannot be countable if it is torsion-free, contradicting that $\text{Hom}(B/U, B/U)$ is a subgroup of the countable torsion-free group $H_B(B/U)$. □

A torsion-free group B is a **finitely faithful S-group** if $r_p(E) = (r_p(B))^2$ for all $p \in \mathbb{P}$. If B is a finitely faithful S-group, then $S_B(G)$ is a pure subgroup of the torsion-free group G by [AlGoe2]. A group B is **faithfully flat** if it is faithful and flat as a left E-module, i.e. the tensor product preserves exact sequences. In particular, if B is faithfully flat, then $T_B(M) = 0$ yields $M = 0$. Every reduced torsion-free group with E hereditary is faithfully flat. Finally, a group G is **locally B-projective** if every finite subset of G is contained in a finitely B-projective summand of B. If B is a torsion-free group of finite rank, then H_B and T_B induce a category equivalence between the categories of locally B-projective groups and locally projective E-modules by [ArMu].

Lemma 1.5.4 *Let B be a faithfully flat group. If G is B-solvable, then the following hold:*

i) *If U is a submodule of $H_B(G)$, and $\theta : T_B(U) \to UB$ is the evaluation map defined by $\theta(u \otimes b) = u(b)$, then θ is an isomorphism.*

ii) *If U and V are submodules of $H_B(G)$ with $UB = VB$ then $U = V$.*

Proof: i) Clearly, θ is onto. To see that it is one-to-one, consider the commutative diagram

$$\begin{array}{ccccc} 0 & \longrightarrow & T_B(U) & \longrightarrow & T_BH_B(G) \\ & & \downarrow \theta & & \downarrow \wr\, \theta_G \\ 0 & \longrightarrow & UB & \longrightarrow & G \end{array}$$

whose top-row is exact since B is flat.

ii) Since $UB = VB = (U+V)B$, it suffices to consider the case that $U \subseteq V$. By i), the evaluation maps $T_B(U) \to UB$ and $T_B(V) \to VB$ are isomorphisms. These maps are the vertical maps in the commutative diagram

$$\begin{array}{ccccccccc} 0 & \longrightarrow & T_B(U) & \longrightarrow & T_B(V) & \longrightarrow & T_B(V/U) & \longrightarrow & 0 \\ & & \downarrow \wr & & \downarrow \wr & & & & \\ 0 & \longrightarrow & UB & = & VB & \longrightarrow & 0. & & \end{array}$$

Thus, $T_B(V/U) = 0$. Since B is faithfully flat, $V/U = 0$. □

Lemma 1.5.5 *Let B be a faithfully flat group of finite rank such that $N(E) = 0$. If G is a torsion-free B-solvable group, and U is a B-generated subgroup of G. Then the purification of U in G is also B-generated.*

Proof: Consider the induced sequence $0 \to H_B(U) \to H_B(G)$ where $H_B(G)$ is a non-singular right E-module. If W is the S-closure of $H_B(U)$ in $H_B(G)$, then $W/H_B(U)$ is torsion and $H_B(G)/W$ is torsion-free. Let $\theta : T_B(W) \to G$ be defined by $\theta(w \otimes a) = w(a)$. Since G is A-solvable, θ is an isomorphism, and $U_* = \theta(T_B(W))$ is B-generated by Lemma 1.5.4. □

1.6 Murley-Groups

In this section we focus on Murley-groups as they will play an important role in our later investigations.

Definition 1.6.1 *A torsion-free group B is called a **Murley-group** if $r_p(B) \leq 1$ for all primes $p \in \mathbb{P}$.*

Furthermore we need

Definition 1.6.2 *Let B be a torsion-free group.*

i) *A subgroup $U \leq B$ is called **fully invariant** if U is an $\operatorname{End}(B)$-submodule of B.*

ii) *B is called **irreducible** if every non-zero pure fully invariant subgroup U of B equals B.*

There are some well-known facts about Murley-groups B:
By [AlGoe] an indecomposable Murley-group is irreducible if and only if it is strongly homogeneous; moreover every irreducible indecomposable Murley-group B is of the form $B_\tau \otimes R$, where B_τ is a rank-1 group of type τ containing B_0 with B_τ/B_0 has bounded p-components and $R = \operatorname{End}(B)$ is a PID, whose additive group R^+ is again an irreducible indecomposable Murley-group of rank $r_0(B)$. We fix this notation for the rest of this section. Furthermore, if V is a non-zero pure subgroup of an irreducible indecomposable Murley-group A, then $r_p(V) = r_p(B)$ for all $p \in \mathbb{P}$ and B/V is divisible.

Irreducible, indecomposable Murley-groups arise in a natural way in algebraic number theory: For example consider the intersection of all localisations of the integral closure by its primeideals.

We now turn our attention to the case, that B is a rank-2 group. In particular, we consider p-local groups B (of rank 2), that is $B_0 = \mathbb{Q}_p$. The next theorem explicitely describes the structure of these groups B:

Theorem 1.6.3 *A torsion-free p-local group B of rank 2 belongs to one of the following five classes of groups:*

i) $B \cong \mathbb{Q} \oplus \mathbb{Q}$;

ii) $B \cong \mathbb{Q}_p \oplus \mathbb{Q}$;

iii) $B \cong \mathbb{Q}_p \oplus \mathbb{Q}_p$;

iv) $E = \mathrm{End}(B) \cong \mathbb{Q}_p$ *and whenever $0 \neq U$ is a $\mathbb{Q}_p$-submodule of B, then U is either isomorphic to $\mathbb{Q}_p$, $\mathbb{Q}_p \oplus \mathbb{Q}_p$ or B; furthermore, if U is pure with $r_0(U) = 1$, then $B/U \cong \mathbb{Q}$, i.e. $r_p(B) = 1$;*

v) *E is an integral domain quasi-isomorphic to B, $r_p(B) = 1$ and if $0 \neq U$ is a $\mathbb{Q}_p$-submodule of B, then U is either isomorphic to $\mathbb{Q}_p$, $\mathbb{Q}_p \oplus \mathbb{Q}_p$ or quasi-equal to B.*

Proof: If B is not reduced, then we trivially are in case i) or ii). So we assume, that B is reduced. Now B is either strongly indecomposable or quasi-equal to $\mathbb{Q}_p \oplus \mathbb{Q}_p$. But in the latter B is a finitely generated $\mathbb{Q}_p$-module and hence a free one, which implies iii).

So let B be strongly indecomposable. If F is a p-basic $\mathbb{Q}_p$-submodule of B, then B/F is torsion-free divisible since $B_0 = \mathbb{Q}_p$. If $r_0(F) = 2$ then we deduce $F = B$ because of the purity, contradicting that B is strongly indecomposable. Since also $F \neq 0$ we directly conclude that $r_0(F) = 1$ and hence $F \cong \mathbb{Q}_p$. This just implies that $r_p(B) = 1$.

Now let U be a $\mathbb{Q}_p$-submodule of B with $r_0(U) = 1$ and U_* its purification. Then $U \cong \mathbb{Q}_p$ and thus $r_p(B/U_*) = 0$, which means that $B/U_* \cong \mathbb{Q}$.
If U is a rank-2 $\mathbb{Q}_p$-submodule of B then $U \supseteq \mathbb{Q}_p \oplus \mathbb{Q}_p = F_1$. If U is not strongly indecomposable, then $U \cong F_1$ and we are done. So let U be strongly indecomposable, which implies that $r_p(U) = 1$. Hence we have $U/F_1 \leq D/F_1 \cong \mathbb{Z}_{p^\infty} \oplus \mathbb{Z}_{p^\infty}$, where

D denotes the divisible hull of F_1. So $U/F_1 \cong \mathbb{Z}_{p^\infty} \oplus T$, where T is cyclic, or U/F_1 is finite, which contradicts the fact that U is strongly indecomposable. We now enlarge F_1 to obtain a free $\mathbb{Q}_p$-module F_2 such that $U/F_2 \cong \mathbb{Z}_{p^\infty}$. Note that F_2 has to be a full-rank free $\mathbb{Q}_p$-submodule of B, hence $U/F_2 \leq B/F_2 \cong \mathbb{Z}_{p^\infty} \oplus T_1$, where T_1 is cyclic. But then B/U is finite and thus B is quasi-equal to U.

If now $r_0(E) = 1$ then $E \cong \mathbb{Q}_p$ and B is a finitely faithful S-group. Hence all finite p-groups T are B-solvable, i.e. the map $\theta_T : \mathrm{Hom}(B,T) \otimes B \to T$ is an isomorphism. We consider the short exact sequence

$$0 \to U \to B \to T \to 0,$$

where T is B-solvable, and obtain the short exact sequence

$$0 \to \mathrm{Hom}(B,U) \to \mathrm{Hom}(B,B) \to \mathrm{Hom}(B,T) \to 0$$

by Lemma 1.2.1. Note that this sequence stays exact if we tensor it by B since B is flat. But then we obtain the commutative diagram

$$\begin{array}{ccccccccc}
0 & \to & \mathrm{Hom}(B,U) \otimes B & \to & \mathrm{Hom}(B,B) \otimes B & \to & \mathrm{Hom}(B,T) \otimes B & \to & 0 \\
& & \uparrow \varphi & & \uparrow \psi_1 & & \uparrow \psi_2 & & \\
0 & \to & U & \to & B & \to & T & \to & 0.
\end{array}$$

Since ψ_1 and ψ_2 are isomophisms we directly conclude that also φ must be an isomorphism. Hence we get $U \cong \mathrm{Hom}(B,U) \otimes B \cong B$.

If otherwise $r_0(E) \geq 2$, then $\mathbb{Q}E$ is a quadratic number field and hence $r_0(E) = 2$. Moreover, E is an integral domain with $r_p(E) = r_p(B) = 1$ since we know that $r_p(E) \leq (r_p(B))^2 = 1$.

So it remains to show that E is quasi-isomorphic to B. Therefore let $0 \neq b \in B$. Since $r_p(E^+) = 1$ the group E^+ must be strongly indecomposable, hence any pure rank-1 image of E^+ is divisible. Considering the short exact sequence

$$0 \to I \to E \to Eb \to 0$$

we see that $I = 0$, because otherwise $E/I \cong Eb$ was divisible, a contradiction. Hence $Eb \cong E$ is an E-module. Now we look at B/Eb and choose, similarly to the above, an F such that $Eb/F \cong \mathbb{Z}_{p^\infty}$. Also as above we obtain that B is a finitely generated E-module and hence B is quasi-isomorphic to E. □

Note that a group B as in case iv) or v) is a Murley-group. In particular, if B is as in case v), it is an irreducible Murley-group since B is quasi-isomorphic to its endomorphism ring.

Now we characterize the quotient groups of $B \oplus B$ modulo pure subgroups.

Theorem 1.6.4 *Let U be a pure subgroup of $B \oplus B$ and let B be as in case iv). Then the following hold:*

i) *If $r_0(U) = 3$ then $(B \oplus B)/U$ is divisible.*

ii) *If $r_0(U) = 2$ then either $(B \oplus B)/U \cong B$ and U is a direct summand of $B \oplus B$ or $(B \oplus B)/U$ is divisible.*

iii) *If $r_0(U) = 1$ then $(B \oplus B)/U$ is a reduced group with $\mathrm{Hom}((B \oplus B)/U, B) = 0$ or $B \oplus B \cong \mathbb{Q} \oplus B$.*

Proof: Choose subgroups V and W of $B \oplus B$ containing U such that W/U is divisible, V/U is reduced, and $(B \oplus B)/U = W/U \oplus V/U$. Then

$$G = (B \oplus B)/W \cong ((B \oplus B)/U)/(W/U) \cong V/U$$

is reduced.

i) If $(B \oplus B)/U$ is not divisible, then it is reduced. Therefore, $(B \oplus B)/U \cong \mathbb{Q}_p$, from which we obtain that $B \oplus B \cong \mathbb{Q}_p \oplus U$ because $\mathrm{Ext}(\mathbb{Q}_p, U) = 0$. Since $E(B) = \mathrm{End}(B)$ is a local ring, every direct summand of $B \oplus B$ is a direct sum of copies of B by Azumaya's Theorem, a contradiction.

ii) If $(B \oplus B)/U$ is not divisible, then V/U has either rank 1 or rank 2. If V/U has rank 1, then it is a reduced rank-1 quotient of $B \oplus B$, which cannot exist by part i). Thus, $G = (B \oplus B)/W$ is a reduced B-generated group of rank 2. Since every rank-1 image of B is divisible, we obtain that every non-zero map $B \to G$ has to be a monomorphism. Suppose that $H \cong B$ is a subgroup of G, and select a full free $\mathbb{Q}_p$-submodule F of H such that $H/F \cong \mathbb{Z}_{p^\infty}$. Since $r_0(F) = r_0(G) = 2$, we can choose an injective hull D of F containing G. Since $D/F \cong \mathbb{Z}_{p^\infty} \oplus \mathbb{Z}_{p^\infty}$, and $H/F \cong \mathbb{Z}_{p^\infty}$, we obtain that G/H has to be finite. Since B is a finitely faithful S-group with $E(B)$ a PID, G/H is B-solvable, and the same holds for the B-generated group G. In particular, $\mathrm{Hom}(B, G)$ is a finitely generated $E(B)$-module of rank 1, and hence $\mathrm{Hom}(B, G) \cong E(B)$. But then, $G \cong \mathrm{Hom}(B, G) \otimes_E B \cong B$. Therefore, we have an exact sequence

$$0 \to U \to B \oplus B \to B \to 0,$$

which splits by Baer's Lemma since $E(B)$ is a PID.

iii) If $(B \oplus B)/U$ is not reduced, then W/U has rank 0, 1, or 2. If $r_0(W/U) = 2$, then V/U is a reduced rank 1 quotient of $B \oplus B$ which cannot exist by i). On the other hand, if $r_0(W/U) = 1$, then $W/U \cong \mathbb{Q}$ and V/U is a reduced rank 2 quotient of $B \oplus B$. By ii), $V/U \cong B$. We consider the short exact sequence

$$0 \to W \to B \oplus B \to B \to 0$$

induced by this isomorphism. By ii), W is a direct summand of $B \oplus B$, say $B \oplus B = W \oplus X$ where $X \cong B$. Then, $(B \oplus B)/U \cong W/U \oplus X \cong \mathbb{Q} \oplus B$.

It remains to consider the case that $(B \oplus B)/U$ is reduced. Since $U \cong \mathbb{Q}_p$, we obtain an exact sequence

$$0 \to \mathbb{Q}_p \to B \oplus B \to (B \oplus B)/U \to 0$$

which induces the sequence

$$0 \to \mathrm{Hom}(B \oplus B)/U, B) \to \mathrm{Hom}(B \oplus B, B) \to M \to 0$$

for some subgroup M of $\mathrm{Hom}(\mathbb{Q}_p, B) \cong B$. Since $\mathrm{Hom}(B \oplus B, B) \cong \mathbb{Q}_p \oplus \mathbb{Q}_p$ is a homogeneous completely decomposable group, the last sequence splits. We obtain the commutative diagram

$$
\begin{array}{ccccccccc}
0 & \to & \mathrm{Hom}(M, B) & \to & \mathrm{Hom}(H_{B \oplus B}(B), B)) & \to & \mathrm{Hom}(H_{(B \oplus B)/U}(B), B) & \to & 0 \\
& & \uparrow \psi & & \uparrow \psi_{B \oplus B} & & \uparrow \psi_{(B \oplus B)/U} & & \\
0 & \to & U & \to & B \oplus B & \to & (B \oplus B)/U & \to & 0.
\end{array}
$$

A simple diagram chase shows that $\psi_{(B \oplus B)/U}$ is onto. Since the top-row of the diagram splits, $\mathrm{Hom}(\mathrm{Hom}((B \oplus B)/U, B), B)$ is B-projective, and thus isomorphic to 0, B, or B^2. A simple rank argument shows that the latter is not possible. On the other hand, if $\mathrm{Hom}((B \oplus B)/U, B) \neq 0$, then the epimorphism $\psi_{(B \oplus B)/U}$ splits, say $(B \oplus B)/U = C \oplus B'$ with $B' \cong B$. In particular, C is a rank-1 quotient of $B \oplus B$ and hence divisible. □

1.7 First results on torsion-free Ext

In this section we summarize some notations and important results that will be used in the reminder of this work.

We begin with two results proven in [Frie].

Theorem 1.7.1 *Let A and B be torsion-free. Then the following hold for the tensor-product $A \otimes B_0$:*

i) $\operatorname{Hom}(A, B) \cong \operatorname{Hom}(A \otimes B_0, B)$;

ii) $\operatorname{Ext}(A, B) \cong \operatorname{Ext}(A \otimes B_0, B)$.

Proof: See [Frie, Thm. 1.5.10]. □

Now we characterize when Ext vanishes in the countable case.

Theorem 1.7.2 *Let A,B be torsion-free, A countable and $|B| < 2^{\aleph_0}$. Then $\operatorname{Ext}(A, B) = 0$ if and only if $A \otimes B_0$ is a free B_0-module.*

Proof: See [Frie, Thm. 3.2.3]. □

The next result helps us to calculate the p-ranks of extension groups. This is useful, because $\operatorname{Ext}(A, B)$ is torsion-free if $r_p(\operatorname{Ext}(A, B)) = 0$ for all primes $p \in \mathbb{P}$.

Lemma 1.7.3 *Let A and B be torsion-free groups of finite rank. Then*

$$r_p(\operatorname{Ext}(A, B)) = r_p(A) \cdot r_p(B) - r_p(\operatorname{Hom}(A, B)).$$

Proof: See [Wa, Thm. 2]. □

We directly conclude a first simple condition for a torsion-free extension group.

Lemma 1.7.4 *Let A and B be torsion-free. Then $\operatorname{Ext}(A, B)$ is torsion-free if A or B is p-divisible for any prime $p \in \mathbb{P}$.*

Proof: See [Frie, Thm. 3.2.1 and Thm 3.4.2]. □

Defining the **support** of a group A as $\operatorname{supp}(\mathbf{A}) = \{\mathbf{p} \in \mathbb{P} \mid \mathbf{pA} \neq \mathbf{A}\}$, that is, the set of all primes not dividing A, there is a simple conclusion which was, although differently formulated, already shown in [Goe, Prop. 1.1]:

Corollary 1.7.5 *If $\operatorname{supp}(A) \cap \operatorname{supp}(B) = \emptyset$ then $\operatorname{Ext}(A, B)$ is torsion-free for arbitrary torsion-free groups A and B. In particular, $\operatorname{Ext}(\mathbb{Q}, B)$ is torsion-free for any torsion-free group B.*

Proof: Clear by the statements above. □

The next theorem will be helpful to investigate the torsion-freeness of extension groups.

Theorem 1.7.6 *Let A' be a pure subgroup of a torsion-free group A. Then $\mathrm{Ext}(A', B)$ is isomorphic to a direct summand of $\mathrm{Ext}(A, B)$ for any group B.*

Proof: At first we consider the short exact sequence

$$0 \to A' \to_* A \to A/A' \to 0.$$

By Lemma 1.2.1 the sequence

$$\cdots \to \mathrm{Ext}(A/A', B) \xrightarrow{\alpha} \mathrm{Ext}(A, B) \to \mathrm{Ext}(A', B) \to 0$$

is exact, too. Since A is torsion-free and A' is pure in A the group A/A' is also torsion-free hence $\mathrm{Ext}(A/A', B)$ is divisible and so is $\mathrm{Im}(\alpha)$. Thus the short exact sequence

$$0 \to \mathrm{Im}(\alpha) \to \mathrm{Ext}(A, B) \to \mathrm{Ext}(A', B) \to 0$$

splits and hence $\mathrm{Ext}(A, B) \cong \mathrm{Ext}(A', B) \oplus \mathrm{Im}(\alpha)$. □

The above theorem fails for subgroups A' of A, which are not pure. For example, choose $A = \mathbb{Q}$, $B = \mathbb{Z}$ and $A' = \mathbb{Q}_p$. Then $\mathrm{Ext}(Q, Z)$ is torsion-free but $\mathrm{Ext}(\mathbb{Q}_p, \mathbb{Z})$ is torsion. Thus it cannot be a direct summand of $\mathrm{Ext}(\mathbb{Q}, \mathbb{Z})$.

Next we present an analogon of Theorem 1.7.6 is

Theorem 1.7.7 *Let A be torsion-free and B' a subgroup of an arbitrary group B. Then $\mathrm{Ext}(A, B/B')$ is isomorphic to a direct summand of $\mathrm{Ext}(A, B)$.*

Proof: Here we have to consider the short exact sequence

$$0 \to B' \to B \to B/B' \to 0.$$

With similar arguments on the exact sequence

$$\cdots \to \mathrm{Ext}(A, B') \xrightarrow{\alpha} \mathrm{Ext}(A, B) \to \mathrm{Ext}(A, B/B') \to 0$$

as in the proof of Theorem 1.7.6 we see that

$$\mathrm{Ext}(A, B) \cong \mathrm{Ext}(A, B/B') \oplus \mathrm{Im}(\alpha).$$

Note that this theorem fails if A is not torsion-free. For example choose $A = \mathbb{Z}_{p^\infty}$, $B = \mathbb{Z}$ and $B' = p\mathbb{Z}$. Then $\mathrm{Ext}(\mathbb{Z}_{p^\infty}, \mathbb{Z})$ is torsion-free, because it is isomorphic to the p-adic completion of J_p, the additive group of the p-adic integers, but $\mathrm{Ext}(\mathbb{Z}_{p^\infty}, \mathbb{Z}/p\mathbb{Z})$ is torsion. Again this means, that $\mathrm{Ext}(\mathbb{Z}_{p^\infty}, \mathbb{Z}/p\mathbb{Z})$ cannot be a direct summand of $\mathrm{Ext}(\mathbb{Z}_{p^\infty}, \mathbb{Z})$.

We immediately derive the following

Corollary 1.7.8 *For any pure subgroup A' of a torsion-free group A and any epimorphic image B' of an arbitrary group B the group $\mathrm{Ext}(A', B')$ is torsion-free if $\mathrm{Ext}(A, B)$ is.*

Proof: This follows directly by Theorems 1.7.6 and 1.7.7. □

Recall the definition of the A-**socle of a group** B, denoted by $S_A(B)$, as the subgroup of B generated by $\mathrm{Hom}(A,B)(A)$ and let $S_A^*(B)$ be its purification in B. Furthermore, define $K_B(A)$ as $\bigcap_{\varphi \in \mathrm{Hom}(A,B)} \mathrm{Ker}(\varphi)$, which is called the B-**radical of** A. With these notation we get a quite useful result for our later investigatons.

Theorem 1.7.9 *Let A and B be torsion-free B_0-modules with $\mathrm{Ext}(A, B)$ torsion-free and B reduced. Then the following hold:*

i) *If $\mathrm{Hom}(A, B) = 0$, then A is divisible.*

ii) *$K_B(A)$ is the largest divisible subgroup of A.*

Proof: i) Let D be the divisible hull of A. We consider the short exact sequence

$$0 \to A \to D \to T \to 0,$$

where T is a divisible torsion group. By Lemma 1.2.1 also the sequence

$$0 = \mathrm{Hom}(A, B) \to \mathrm{Ext}(T, B) \to \mathrm{Ext}(D, B) \to \mathrm{Ext}(A, B) \to 0$$

is exact. Since $\mathrm{Ext}(A, B)$ is torsion-free the group $\mathrm{Ext}(T, B)$ has to be divisible as a pure subgroup of the torsion-free group $\mathrm{Ext}(D, B)$. But $\mathrm{Ext}(T, B)$ is also reduced by [Fu1, Theorem 52.3], so that we obtain $\mathrm{Ext}(T, B) \cong \mathrm{Hom}(T, \mathbb{Q}B/B) = 0$. But this is only possible, if $T_p = 0$ whenever $(\mathbb{Q}B/B)_p \neq 0$. Thus $T_p \neq 0$ yields $B = pB$.

Since A is a B_0-module, also $A = pA$. But then $T_p = 0$ and A is divisible.

ii) We consider the short exact sequence
$$0 \to K_B(A) \to A \to A/K_B(A) \to 0,$$
which induces the exact sequence
$$0 \to \operatorname{Hom}(A/K_B(A), B) \xrightarrow{\alpha} \operatorname{Hom}(A, B) \to \operatorname{Hom}(K_B(A), B)$$
$$\xrightarrow{\delta} \operatorname{Ext}(A/K_B(A), B) \to \operatorname{Ext}(A, B) \to \operatorname{Ext}(K_B(A), B) \to 0.$$

Since α is an isomorphism the map δ must be a monomorphism and hence $\operatorname{Hom}(K_B(A), B)$ is isomorphic to a pure subgroup of the divisible group $\operatorname{Ext}(A/K_B(A), B)$, so that it is divisible itself. However, dual groups are reduced, so $\operatorname{Hom}(K_B(A), B) = 0$. But then the sequence
$$0 \to \operatorname{Ext}(A/K_B(A), B) \to \operatorname{Ext}(A, B) \to \operatorname{Ext}(K_B(A), B) \to 0$$
is exact and splits. Thus $\operatorname{Ext}(K_B(A), B)$ is torsion-free and $K_B(A)$ is divisible by part i). Now any divisible subgroup of A is mapped onto 0 since B is reduced. So $K_B(A)$ is the largest divisible subgroup of A. □

The central tool in the investigation of torsion-free Ext is the following result of Goeters:

Theorem 1.7.10 *Let A and B be countable torsion-free groups with B of finite rank. If $\operatorname{Ext}(A, B)$ is torsion-free and $OT(B) \neq tp(\mathbb{Q})$ then B is injective with respect to*
$$0 \to H \to_* A \to A/H \to 0$$
for all finite rank pure subgroups H of A.
Dually, A is projective with respect to
$$0 \to H \to_* B \to B/H \to 0$$
for any (finite rank) pure subgroup H of B.

Proof: See [Goe, Theorem 2.2]. □

Unfortunately, it remains open to what extent the last result remains true in the important case that $OT(B) = tp(\mathbb{Q})$.

We extend the property being "injective with resepect to" given in Theorem 1.7.10 to the notion B-cobalanced:

Definition 1.7.11 *Let A be a torsion-free group and A' be a pure subgroup of A. Then we call the short exact sequence*

$$0 \to A' \to_* A \to A/A' \to 0$$

*B-**cobalanced** if the induced sequence*

$$\operatorname{Hom}(A, B) \to \operatorname{Hom}(A', B) \to 0$$

is exact.

Now we want to replace the pure subgroups A' of a torsion-free group A in Corollary 1.7.8 by factor groups of A.

Theorem 1.7.12 *Let A' be a finite rank pure subgroup of a countable torsion-free group A and let B be a finite rank group with $OT(B) \neq tp(\mathbb{Q})$ and torsion-free $\operatorname{Ext}(A, B)$. Furthermore let B' be a pure subgroup of B. Then also $\operatorname{Ext}(A/A', B)$ and $\operatorname{Ext}(A, B')$ are torsion-free.*

Proof: By Theorem 1.7.10 B is injective with respect to

$$0 \to A' \to_* A \to A/A' \to 0.$$

This means that any $\varphi \in \operatorname{Hom}(A', B)$ extends to a $\tilde{\varphi} \in \operatorname{Hom}(A, B)$. Hence the map β in the exact sequence

$$0 \to \operatorname{Hom}(A/A', B) \to \operatorname{Hom}(A, B) \xrightarrow{\beta} \operatorname{Hom}(A', B)$$

$$\to \operatorname{Ext}(A/A', B) \xrightarrow{\alpha} \operatorname{Ext}(A, B) \to \operatorname{Ext}(A', B) \to 0$$

is onto. So we have the short exact sequence

$$0 \to \operatorname{Ext}(A/A', B) \xrightarrow{\alpha} \operatorname{Ext}(A, B) \to \operatorname{Ext}(A', B) \to 0$$

and thus $\operatorname{Im}(\alpha) \cong \operatorname{Ext}(A/A', B)$. Since $\operatorname{Im}(\alpha)$ is torsion-free the group $\operatorname{Ext}(A/A', B)$ is torsion-free, too.

On the other hand, A is projective with respect to

$$0 \to B' \to_* B \to B/B' \to 0$$

which means that any element of $\operatorname{Hom}(A, B/B')$ extends to an element of $\operatorname{Hom}(A, B)$ and thus implies that we have a short exact sequence

$$0 \to \operatorname{Ext}(A, B') \xrightarrow{\alpha} \operatorname{Ext}(A, B) \to \operatorname{Ext}(A, B/B') \to 0.$$

Since $\operatorname{Ext}(A, B)$ is torsion-free we conclude $\operatorname{Ext}(A, B')$ is also torsion-free. □

Note that the condition of purity is in both components necessary. For example, we consider on the one hand the group $\operatorname{Ext}(\mathbb{Q}_p, \mathbb{Q}^p)$ which is trivially torsion-free. The subgroup $\mathbb{Z}$ is not pure in $\mathbb{Q}^p$ and $\operatorname{Ext}(\mathbb{Q}_p, \mathbb{Z})$ is not torsion-free. On the other hand, the group $\operatorname{Ext}(\mathbb{Z}, \mathbb{Z})$ is torsion-free and $p\mathbb{Z}$ is not pure in $\mathbb{Z}$. Here we have $\operatorname{Ext}(\mathbb{Z}/p\mathbb{Z}, \mathbb{Z}) \cong \mathbb{Z}/p\mathbb{Z}$ and this is not torsion-free.

The following nice result is based on an idea of Pat Goeters. It naturally extends the well-known theorem by Pontryagin that a countable torsion-free group is free if and only if all its pure subgroups of finite rank are free.

Theorem 1.7.13 *Let A and B be countable torsion-free groups and let B be of finite rank with $OT(B) \neq tp(\mathbb{Q})$. Then $\operatorname{Ext}(A, B)$ is torsion-free if and only if $\operatorname{Ext}(A', B)$ is torsion-free for all finite rank pure subgroups A' of A.*

Proof: See [Goe, Theorem 3.1]. □

We finish this section with another important result of P. Goeters.

Theorem 1.7.14 *For torsion-free groups A and B the following are equivalent:*

i) $\operatorname{Ext}(A, B)$ *is torsion-free;*

ii) $\operatorname{Ext}(A, S_A^*(B))$ *is torsion-free and* $\operatorname{supp}(A) \cap \operatorname{supp}(B/S_A^*(B)) = \emptyset$;

iii) $\operatorname{Ext}(A/K_B(A), B)$ *is torsion-free and* $\operatorname{supp}(K_B(A)) \cap \operatorname{supp}(B) = \emptyset$.

Proof: See [Goe, Theorem 1.2]. □

Chapter 2

The Finite Rank Case

After we presented some fundamental background in the previous chapter, we are now able to start our investigations of torsion-free extensions of torsion-free groups of finite rank.

2.1 On a Result of Goeters

This section is motivated by the following two results of Pat Goeters. At first we have

Lemma 2.1.1 *Let A and B be of finite rank and $OT(A) \leq IT(B)$. Then $\text{Ext}(A, B)$ is torsion-free.*

Proof: See [Goe, Prop. 1.7]. □

Furthermore we know

Proposition 2.1.2 *Let A and B be of finite rank and $OT(B) \neq tp(\mathbb{Q})$. Then the following are equivalent:*

  i) $\text{Ext}(A, B)$ *is torsion-free;*

  ii) $K_B(A)$ *is p-divisible for all $p \in \text{supp}(B)$ and $OT(A/K_B(A)) \leq IT(B)$.*

Proof: See [Goe, Theorem 1.6]. □

To extend these two results we first consider the case that A is p-reduced.

Theorem 2.1.3 *Let A and B be of finite rank and let $p \in \mathbb{P}$ such that A is p-reduced and B is not p-divisible with $OT(B) \neq tp(\mathbb{Q})$. Then the following are equivalent:*

i) $\text{Ext}(A, B)$ *is torsion-free;*

ii) $OT(A) \leq IT(B)$.

Proof: First let $K_B(A) \neq 0$, so there is a non-zero $a \in K_B(A)$. W.l.o.g. we may assume that $a \notin pA$, because otherwise we have $a = pa'$ and thus $a' \in K_B(A)$. Since A is p-reduced this procedure must end. So $[a] \not\equiv 0 \pmod{pA}$. Furthermore, A/pA and B/pB are $\mathbb{Z}/p\mathbb{Z}$ vector-spaces. So there exists a $\varphi \in \text{Hom}(A/pA, B/pB)$ with $\varphi([a]) \neq 0$. But $a \in K_B(A)$, so φ has no lifting to an element of $\text{Hom}(A, B)$. Therefore $\text{Ext}(A, B)$ cannot be torsion-free. The contraposition is, that $K_B(A) = 0$ if $\text{Ext}(A, B)$ is torsion-free. So the implication i) $\Rightarrow$ ii) is clear by Prop. 2.1.2. The other direction was done in Lemma 2.1.1. $\square$

We can remove the condition on $K_B(A)$ from Proposition 2.1.2:

Theorem 2.1.4 *Let A and B be torsion-free groups of finite rank and $OT(B) \neq tp(\mathbb{Q})$. Then $\text{Ext}(A, B)$ is torsion-free if and only if*

$$OT((A \otimes B_0)/D) \leq IT(B),$$

where D is the divisible subgroup of $A \otimes B_0$.

Proof: Recall that we have

$$\text{Ext}_{\mathbb{Z}}(A, B) \cong \text{Ext}_{B_0}(A \otimes B_0, B),$$

so w.l.o.g. A is a B_0-module with $A_0 \geq B_0$. Hence also $K_B(A \otimes B_0)$ is a B_0-module and thus p-divisible for all primes $p \notin \text{supp}(B)$. If now $\text{Ext}(A, B)$ is torsion-free, then $K_B(A \otimes B_0)$ is also p-divisible for all primes $p \in \text{supp}(B)$ by Lemma 2.1.2. So $K_B(A \otimes B_0)$ is divisible and this means that $K_B(A \otimes B_0) \leq D$ where D is the divisible subgroup of $A \otimes B_0$ and therefore we have $OT((A \otimes B_0)/D) \leq IT(B)$.

To prove the other implication let A be such that $OT((A \otimes B_0)/D) \leq IT(B)$. Then the group $\text{Ext}((A \otimes B_0)/D, B)$ is torsion-free by Proposition 2.1.1. Since D has to be a direct summand we directly conclude that also the group $\text{Ext}(A \otimes B_0, B)$ is torsion-free and hence $\text{Ext}(A, B)$ is. $\square$

In this notation we suppose that $OT(0) \leq IT(B)$ for any finite rank group B. So the case of divisible $A \otimes B_0$ is included.

Note that this result fails if $OT(B) = tp(\mathbb{Q})$. For example we consider the group $B = \mathbb{Q}_p \oplus \mathbb{Q}_q$ for distinct primes p and q. Then we have

$$\operatorname{Ext}(\mathbb{Q}_p, B) \cong \operatorname{Ext}(\mathbb{Q}_p, \mathbb{Q}_p) \oplus \operatorname{Ext}(\mathbb{Q}_p, \mathbb{Q}_q),$$

which is torsion-free. But $OT((\mathbb{Q}_p \otimes B_0)/D) = tp(\mathbb{Q}_p) > IT(B)$.

We directly derive the following reflection on the nucleus B_0 of a group B.

Corollary 2.1.5 *Let A and B be torsion-free groups of finite rank. Then $\operatorname{Ext}(A, B)$ is torsion-free whenever $\operatorname{Ext}(A, B_0)$ is.*

Proof: In the first case let $B_0 = \mathbb{Q}$. Then B is divisible and hence $\operatorname{Ext}(A, B)$ is torsion-free.
Now let $B_0 \neq \mathbb{Q}$. Then we may apply Theorem 2.1.4 to see that

$$OT((A \otimes B_0)/D) \leq tp(B_0).$$

But we also have $tp(B_0) \leq IT(B)$. So we immediately conclude that $\operatorname{Ext}((A \otimes B_0)/D, B)$ must be torsion-free by Lemma 2.1.1. Since D is a direct summand we deduce that $\operatorname{Ext}(A \otimes B_0, B)$ is torsion-free and hence also $\operatorname{Ext}(A, B)$. $\square$

However, the converse fails in general as the next example shows.

Example 2.1.6 *Let A be a rank-1 group of type $(1, 1, \ldots, 1)$. Then $\operatorname{Ext}(A, A)$ is torsion-free by Proposition 2.1.8. Since $A_0 = \mathbb{Z}$ we have*

$$r_p(\operatorname{Ext}(A, A_0)) = r_p(\operatorname{Ext}(A, \mathbb{Z})) = r_p(A) \cdot r_p(\mathbb{Z}) - r_p(\operatorname{Hom}(A, \mathbb{Z})) = 1$$

for any $p \in \mathbb{P}$ hence $\operatorname{Ext}(A, A_0)$ is not torsion-free.

Now we turn to the concept of torsion-free splitters. P. Schultz called a group G satisfying $\operatorname{Ext}(G, G) = 0$ a **splitter**. Similarly, we have

Definition 2.1.7 *Let A be a group such that $\operatorname{Ext}(A, A)$ is torsion-free. Then we call A a **torsion-free splitter**.*

A direct consequence is

Proposition 2.1.8 *Any rational group R is a torsion-free splitter.*

Proof: Since $OT(R) = IT(R)$ for any rational group R we conclude that $\mathrm{Ext}(R,R)$ is torsion-free by Lemma 2.1.1. $\square$

The next lemma is extracted from an article by R.B. Warfield Jr.

Lemma 2.1.9 *Let A be a finite rank torsion-free group. Then the following are equivalent:*

i) $r_0(\mathrm{Hom}(A,A)) = (r_0(A))^2$;

ii) $IT(A) = OT(A)$;

iii) $A \cong R^n$ for some $n \in \mathbb{N}$ and a rank-1 group R.

Proof: See [Wa1]. $\square$

We are now able to characterize the torsion-free splitters in the class $\mathfrak{Tff}$ of all torsion-free abelian groups of finite rank.

Theorem 2.1.10 *Let $A \in \mathfrak{Tff}$ be a torsion-free splitter and $OT(A) \neq tp(\mathbb{Q})$. Then A is of the form $A \cong R^n$ for some $n \in \mathbb{N}$ and a rank-1 group R.*

Proof: Since A is a torsion-free-splitter we have $\mathrm{Ext}(A,A)$ torsion-free. Furthermore we have $id_A \in Hom(A,A)$ so it is $K_A(A) = 0$ and hence $OT(A) \leq IT(A)$ by Lemma 2.1.2. But in general we have $OT(A) \geq IT(A)$ and thus we conclude $IT(A) = OT(A)$. By Lemma 2.1.9 we directly derive that A must be isomorphic to R^n for some $n \in \mathbb{N}$ and a rank-1 group R. $\square$

Note that Theorem 2.1.10 fails if we remove the condition $OT(A) \neq tp(\mathbb{Q})$. For example let $A = \mathbb{Z} \oplus \mathbb{Q}$. Then we have $OT(A) = tp(\mathbb{Q})$ and $\mathrm{Ext}(A,A) = \mathrm{Ext}(\mathbb{Q}, \mathbb{Z})$ which is torsion-free. So A is a torsion-free splitter but $A \not\cong R^n$ for any rational group R.

Theorem 2.1.11 *Let $B \in \mathfrak{Tff}$ be a torsion-free splitter and $OT(B) \neq tp(\mathbb{Q})$. Then $\mathrm{Ext}(A,B)$ is torsion-free if and only if $K_B((A \otimes B_0)/D) = 0$, where D denotes the maximal divisible subgroup of $A \otimes B_0$.*

Proof: First let $\text{Ext}(A, B) \cong \text{Ext}(A \otimes B_0, B)$ be torsion-free. By Theorem 1.7.9 we know that $K_B(A \otimes B_0) = D$. Hence we obtain that

$$K_B((A \otimes B_0)/D) = 0.$$

Now let $K_B((A \otimes B_0)/D) = 0$, which implies, that $G = (A \otimes B_0)/D \leq \bigoplus_m B$ for some $m \in \mathbb{N}$ by [HuWa]. Since B is a torsion-free-splitter with $OT(B) \neq tp(\mathbb{Q})$, we know that B is of the form $B = R^n$ for a rank-1 group A and $n \in \mathbb{N}$. Now let U be a pure subgroup of G such that $r_0(G/U) = 1$ and U_*, G_* their purifications in B. We consider the map $\varphi : G/U \to G_*/U_*$ with $\varphi(g + U) = g + U_*$. If $g + U_* = 0$, then we have $g \in U_*$ and this implies that there exists a non-zero $k \in \mathbb{Z}$ such that $k \cdot g \in U \leq_* G$. Because of purity we obtain that $g \in U$. Hence φ is a monomorphism and thus $tp(G/U) \leq tp(G_*/U_*)$ by [Ar, Prop. 1.2]. But $\text{Ext}(G_*, B)$ is torsion-free and hence $OT(G_*) \leq IT(B) = tp(R)$ by Lemma 2.1.2 since $K_B(G) = 0$. Furthermore we have $tp(G_*/U_*) \leq OT(G_*)$ by [Ar, Prop. 1.8]. So alltogether we get the inequality

$$tp(G/U) \leq tp(G_*/U_*) \leq OT(G_*) \leq IT(B) = tp(R)$$

and this means that $\text{Ext}(G, B)$ is torsion-free. Moreover $\text{Ext}(D, B)$ is trivially torsion-free, hence $\text{Ext}(A \otimes B_0, B) \cong \text{Ext}(A, B)$ is torsion-free, too. $\square$

Note that this theorem fails, if B is not a torsion-free splitter. For example we consider the rank-1 groups G_1 with $tp(G_1) = (0, 0, 0, \infty, \infty, \ldots)$ and G_2 with $tp(G_2) = (0, 0, \infty, \infty, \ldots)$ and define

$$A = \left\langle G_1 \times G_2, \frac{(1,1)}{p_2^n} \mid n < \omega \right\rangle \leq \mathbb{Q} \oplus \mathbb{Q}.$$

Then A is a strongly indecomposable group of rank 2 with $\text{End}(A) \leq \mathbb{Q}$ by [Ar]. Furthermore it is $OT(A) \neq tp(\mathbb{Q})$ and $IT(A) = tp(G_1)$. If G_3 denotes the purified subgroup of A generated by all elements $\frac{(1,1)}{p_2^n}$ with $n < \omega$ we directly see, that $\text{Ext}(G_3, A)$ cannot be torsion-free. Since $G_3 \leq_* A$ also $\text{Ext}(A, A)$ is not torsion-free but $K_A(A) = 0$.

The main result of this section is

Theorem 2.1.12 *Let B be an indecomposable torsion-free group of finite rank and $OT(B) \neq tp(\mathbb{Q})$. Then the following hold:*

i) *The following are equivalent:*

   a) $r_0(B) = 1$;

   b) *if $A \leq B$ and $\mathrm{Ext}(H, B)$ is torsion-free then a group G fitting into a short pure-exact sequence*
   $$0 \to A \to_* G \to H \to 0$$
   satisfies $\mathrm{Ext}(G, B)$ is torsion-free if and only if
   $r_0(\mathrm{Hom}(G, B)) > r_0(\mathrm{Hom}(H, B))$.

ii) *If $r_0(B) = 1$ then B has idempotent type if and only if every group G fitting into an short pure-exact sequence*
$$0 \to B \to_* G \to B \to 0$$
satisfies $\mathrm{Ext}(G, B)$ is torsion-free.

iii) *[$V = L$] If $B \neq \mathbb{Q}$ is a subring of $\mathbb{Q}$ and H is uncountable then there exists a short pure-exact sequence*
$$0 \to B \to_* G \to H \to 0$$
such that $\mathrm{Ext}(H, G)$ is torsion-free but $\mathrm{Ext}(G, B)$ is not torsion-free.

Proof: i) a) $\Rightarrow$ b) We consider the by Lemma 1.2.1 induced exact sequence
$$0 \to \mathrm{Hom}(H, B) \to \mathrm{Hom}(G, B) \xrightarrow{\alpha} \mathrm{Hom}(A, B)$$
$$\xrightarrow{\delta} \mathrm{Ext}(H, B) \to \mathrm{Ext}(G, B) \to \mathrm{Ext}(A, B) \to 0,$$
in which $\mathrm{Ext}(H, B)$ and $\mathrm{Ext}(A, B)$ are torsion-free since $tp(A) \leq tp(B)$. Therefore $M = \mathrm{Im}(\alpha)$ is a pure subgroup of the rank-1 group $\mathrm{Hom}(A, B)$ because $\mathrm{Hom}(A, B)/M$ is isomorphic to $\mathrm{Im}(\delta)$, which is torsion-free. Thus we either have $M = 0$ or $M = \mathrm{Hom}(A, B)$ and
$$r_0(\mathrm{Hom}(G, B)) = r_0(\mathrm{Hom}(H, B)) + r_0(M).$$

Now suppose that $\mathrm{Ext}(G, B)$ is torsion-free. Then $\mathrm{Im}(\delta)$ is a pure subgroup of $\mathrm{Ext}(H, B)$ and thus divisible. However, this is only possible if $\mathrm{Im}(\delta) = 0$, which means that $r_0(M) = 1$ and hence $r_0(\mathrm{Hom}(G, B)) > r_0(\mathrm{Hom}(H, B))$.

Conversely, suppose that $r_0(\mathrm{Hom}(G, B)) > r_0(\mathrm{Hom}(H, B))$. Then $r_0(M) \neq 0$ and hence $\mathrm{Im}(\delta) = 0$ since $r_0(\mathrm{Hom}(A, B)) = 1$. Thus the sequence
$$0 \to \mathrm{Ext}(H, B) \to \mathrm{Ext}(G, B) \to \mathrm{Ext}(A, B) \to 0$$

is split-exact and $\mathrm{Ext}(G,B)$ has to be torsion-free.

b) $\Rightarrow$ a) We consider the short exact sequence

$$0 \to B \to B \to 0 \to 0,$$

in which $r_0(\mathrm{Hom}(B,B)) > 0$. By b) the group $\mathrm{Ext}(B,B)$ is torsion-free. So B is a homogeneous completely decomposable group by Theorem 2.1.10. Since B is also indecomposable, it must be a rank-1 group.

ii) If B has idempotent type, then G is a B-module and the sequence

$$0 \to B \to G \to B \to 0$$

is split-exact. So $\mathrm{Ext}(G,B)$ trivially has to be torsion-free.

Conversely, we show that $\mathrm{Ext}(B,B) = 0$, which directly implies that B has idempotent type. Therefore we consider the short exact sequence

$$0 \to B \to G \to B \to 0$$

with $\mathrm{Ext}(G,B)$ torsion-free. By i) we know that

$$r_0(\mathrm{Hom}(G,B)) > r_0(\mathrm{Hom}(B,B)) = 1.$$

But then the sequence

$$0 \to \mathrm{Hom}(B,B) \to \mathrm{Hom}(G,B) \to \mathrm{Hom}(B,B) \to 0$$

is exact and has to split. Then also the sequence

$$0 \to B \to G \to B \to 0$$

splits by Lemma 1.1.8 and hence $\mathrm{Ext}(B,B) = 0$. Since B is a rank-1 group we directly conclude that $B = B_0$ by Theorem 1.7.2 and hence B has idempotent type.

iii) Since B is a subring of $\mathbb{Q}$ choose an uncountable coseparable B-module H which is not a B-Whitehead module, i.e. $\mathrm{Ext}(H,B) \neq 0$ and consider a non-splitting short exact sequence

$$0 \to B \to G \to H \to 0.$$

As before we obtain the induced sequence

$$0 \to \mathrm{Hom}(H,B) \to \mathrm{Hom}(G,B) \xrightarrow{\alpha} \mathrm{Hom}(B,B)$$

$$\xrightarrow{\delta} \mathrm{Ext}(H,B) \to \mathrm{Ext}(G,B) \to \mathrm{Ext}(B,B) \to 0.$$

If $\mathrm{Ext}(G, B)$ were torsion-free then $\mathrm{Im}(\delta)$ would be divisible, which can only happen, if α is onto. But then α splits and so does the sequence

$$0 \to B \to G \to H \to 0$$

by Lemma 1.1.8, a contradiction. $\square$

2.2 The Divisible Outer-type Case

As we have seen, most of the previous results fail if $OT(B) = tp(\mathbb{Q})$. Theorem 2.1.10 shows, that this is the more interesting and complex case. Due to the complexity of this case, we consider it seperately. If nothing else is said, all groups are assumed to have finite rank.

Murley-groups arise naturally in the investigations of torsion-free splitters as shown in

Proposition 2.2.1 *Let B be a strongly indecomposable group of rank 2. Then the following are equivalent:*

i) *B is a torsion-free splitter;*

ii) *$OT(B) = tp(\mathbb{Q})$;*

iii) *B is a Murley-group.*

Proof: i) $\Rightarrow$ ii) Since $\mathrm{Ext}(B, B)$ is torsion-free we conclude $r_p(\mathrm{End}(B)) = (r_p(B))^2$ for all $p \in \mathbb{P}$. If $r_p(B) = 2$, then $r_p(\mathrm{End}(B)) = 4$. But $r_p(\mathrm{End}(B)) \le r_0(\mathrm{End}(B))$ so that $r_0(\mathrm{End}(B)) = (r_0(B))^2$, which implies that B is completely decomposable by Lemma 2.1.9, a contradiction. Hence $r_p(B) \le 1 < 2 = r_0(B)$ for each prime p and this means nothing else but $OT(B) = tp(\mathbb{Q})$.

ii) $\Rightarrow$ iii) This is trivial since $OT(B) = tp(\mathbb{Q})$ implies $r_p(B) < r_0(B) = 2$ for all $p \in \mathbb{P}$.

iii) $\Rightarrow$ i) Since B is a Murley-group we have $r_p(B) \le 1$. But then we directly conclude that $r_p(\mathrm{End}(B)) = (r_p(B))^2$ since 0 and 1 are idempotent. So $\mathrm{Ext}(B, B)$ is torsion-free. $\square$

Our next theorem characterizes the irreducible indecomposable Murley-groups.

Theorem 2.2.2 *The following conditions are equivalent for a reduced torsion-free group B of finite rank:*

i) *B is an irreducible indecomposable Murley group.*

ii) *The following hold for B:*

    a) *B is homogeneous and $r_0(E = \mathrm{End}(B)) \leq r_0(B)$;*

    b) *B is a torsion-free splitter;*

    c) *Let A be a torsion-free group with $\mathrm{Ext}(A,B)$ torsion-free. If U is a finite rank pure subgroup of A, then U is B-cobalanced in A.*

iii) *The following hold for B:*

    a) *B is homogeneous;*

    b) *B/U is divisibe for all non-zero pure subgroups U of B;*

    c) *Let A be a torsion-free group with $\mathrm{Ext}(A,B)$ torsion-free. If U is a finite rank pure subgroup of A, then U is B-cobalanced in A.*

Proof: i) $\Rightarrow$ ii) Let B be an irreducible indecomposable Murley-group. Then, trivially, B is homogeneous by the remarks above and of the form $B = B_\tau \otimes R$, so B is homogeneous of type τ and $r_0(R = \mathrm{End}(B)) \leq r_0(B)$. Since clearly $r_p(B) \leq 1$, we have B is a torsion-free splitter by Lemma 1.7.3. Now let $\mathrm{Ext}(A,B)$ be torsion-free; we may assume that A is reduced. Hence $K_B(A) = 0$. Moreover, let

$$0 \to U \to A \to H \to 0$$

be a pure-exact short sequence. We induct on $n = r_0(U)$ and start with $r_0(U) = 1$. By Lemma 1.2.1 we obtain the exact sequence

$$0 \to \mathrm{Hom}(H,B) \to \mathrm{Hom}(A,B) \overset{\alpha}{\to} \mathrm{Hom}(U,B)$$

$$\overset{\delta}{\to} \mathrm{Ext}(H,B) \overset{\alpha}{\to} \mathrm{Ext}(A,B) \to \mathrm{Ext}(U,B) \to 0.$$

Let $M = \mathrm{Im}(\alpha)$. Then $M \neq 0$, because otherwise we would have $U \leq K_B(A) = 0$, a contradiction. Since $\mathrm{Ext}(A,B)$ is torsion-free the group $\mathrm{Hom}(U,B)/M \cong \mathrm{Im}(\delta)$ is a pure subgroup of the divisible group $\mathrm{Ext}(H,B)$ and hence itself divisible. Note that also $\mathrm{Hom}(U,B) \neq 0$, because otherwise also $M = 0$, which leads to a contradiciton as we have seen before. Since $r_0(U) = 1$ we have $tp(U) \leq \tau$ because B is homogeneous

of type τ. Now we choose a subgroup $B' = \bigoplus_{i=1}^{m} B_i \leq B$ such that $m = r_0(B)$ and $B_i \cong B_\tau$. Then
$$0 \to \operatorname{Hom}(U, B') \to \operatorname{Hom}(U, B)$$
is exact and $r_0(B) = r_0(\operatorname{Hom}(U, B')) \leq r_0(\operatorname{Hom}(U, B))$. But on the other hand $r_0(\operatorname{Hom}(U, B)) \leq r_0(U) \cdot r_0(B) = r_0(B)$, so $r_0(\operatorname{Hom}(U, B)) = r_0(B) = r_0(R)$. Now pick $0 \neq u \in \operatorname{Hom}(U, B)$. Then $Ru \leq \operatorname{Hom}(U, B)$ and $Ru \cong R$, so that $\operatorname{Hom}(U, B)$ is a rank-1 R-module. Thus $\operatorname{Hom}(U, B)/M$ is a divisible torsion group. Since $\mathbb{Z} \leq U$ we have an exact sequence
$$0 = \operatorname{Hom}(U/\mathbb{Z}, B) \to \operatorname{Hom}(U, B) \to \operatorname{Hom}(\mathbb{Z}, B) \cong B,$$
so that $\operatorname{Hom}(U, B)$ is an R-submodule of B and the same holds for M. Since B is a rank-1 R-module, B/M and $B/\operatorname{Hom}(U, B)$ are torsion. Now we consider $(M + R)/M \cong R/(M \cap R)$. Since $(M + R)/M \leq B/M$, also $R/(M \cap R)$ is torsion and we can find $0 \neq l$ such that $lR \leq M \cap R$, so that $(M + R)/M$ has bounded p-components for all primes $p \in \mathbb{P}$. Consider
$$B_\tau/B_0 \otimes R \cong B/R \to B/(M + R) \to 0,$$
in which the p-components are maped on the p-components. Thus $B/(M + R)$ has finite p-components, because $B_\tau/B_0 \otimes R$ has. But we also have a short exact sequence
$$0 \to (M + R)/M \to B/M \to B/(M + R) \to 0,$$
in which both of the outside groups have bounded p-components and thus the same holds for the group in the center. But if $\operatorname{Hom}(U, B)/M \leq B/M$ has bounded p-components, this contradicts the fact that $\operatorname{Hom}(U, B)/M$ is divisible unless $\operatorname{Hom}(U, B) = M$. So U is B-cobalanced in A.

Now let $r_0(U) = n$ and $\varphi : U \to B$. Select a pure rank-1 subgroup W of U and consider $\varphi \upharpoonright W$. There is a $\psi : A \to B$ with $\psi \upharpoonright W = \varphi \upharpoonright W$. The map $\lambda = \psi - \varphi \upharpoonright W$ satisfies $\lambda \upharpoonright W = 0$, so that $\lambda \in \operatorname{Hom}(U/W, B)$. By induction hypothesis and $n = 1$, there exists $\beta \in \operatorname{Hom}(A/W, B)$ such that $\beta \upharpoonright U/W = \lambda$. We view $\beta : A \to B$ with $\beta \upharpoonright W = 0$.

Now we consider $\beta + \psi$. For $u \in U$ we have
$$(\beta + \psi)(u) = (\varphi - \psi)(u) + \psi(u) = \varphi(u).$$
Thus U is a B-cobalanced subgroup of A.

ii) $\Rightarrow$ iii) Let A be a pure rank-1 subgroup of B, and consider the induced sequence
$$0 \to \operatorname{Hom}(B/A, A) \to \operatorname{Hom}(B, B) \to \operatorname{Hom}(A, B) \to 0,$$

which is exact by c). From this we obtain $r_0(E) \geq r_0(\mathrm{Hom}(A,B))$. However, by a), B contains a subgroup isomorphic to A^n where $n = r_0(B)$, so that

$$r_0(\mathrm{Hom}(A,B)) \geq r_0(\mathrm{Hom}(A,A^n)) = n = r_0(B).$$

Since $r_0(B) \geq r_0(E)$ by a), we have

$$r_0(E) \geq r_0(\mathrm{Hom}(A,B)) \geq r_0(B) \geq r_0(E)$$

and all these ranks coincide. Thus, the exactness of

$$0 \to \mathrm{Hom}(B/A,A) \to \mathrm{Hom}(B,B) \to \mathrm{Hom}(A,B) \to 0$$

implies $\mathrm{Hom}(B/A,B) = 0$. The same sequence also provides the exactness of

$$0 \to \mathrm{Ext}(B/A,B) \to \mathrm{Ext}(B,B)$$

from which we get that $\mathrm{Ext}(B/A,B)$ is torsion-free. Because $\mathrm{Hom}(B/A,B) = 0$, B/A is divisible by Theorem 1.7.9. If U is an arbitrary pure non-zero subgroup of B then U contains a rank-1 subgroup A, and B/U is divisible as an epimorphic image of B/A.

iii) $\Rightarrow$ i) Let A be a pure rank-1 subgroup of B. Since $r_p(B) = r_p(A) + r_p(B/A) \leq 1$ for all primes p, the group B is a Murley-group. If B is not indecomposable, then $B = C \oplus D$, and one of the summands has to be divisible, which is not possible since B is homogeneous.

It remains to show that B is irreducible. Since B is homogeneous of type τ, we have a short exact sequence

$$0 \to B_\tau \to B \to D \to 0,$$

where D is torsion-free divisible. It induces the exact sequence

$$0 = \mathrm{Hom}(D,B) \to \mathrm{Hom}(B,B) \to \mathrm{Hom}(B_\tau,B) \to 0,$$

since B_τ is B-cobalanced in B. Thus $r_0(R) = r_0(\mathrm{Hom}(B_\tau,B))$. Moreover, we can find a short exact sequence

$$0 \to B' = \bigoplus_{i=1}^{m} B_i \to B \to T \to 0,$$

where $n = r_0(B)$, each $B_i \cong B_\tau$ and T is torsion. It induces

$$0 \to \mathrm{Hom}(B_\tau,B') \to \mathrm{Hom}(B_\tau,B),$$

so that $r_0(\mathrm{Hom}(B_\tau, B)) \geq r_0(\mathrm{Hom}(B_\tau, B')) = r_0(B)$. On the other hand, if we take $0 \neq b \in B$, then Rb is a torsion-free module over the PID R. Thus $Rb \cong R$ and $r_0(R) \leq r_0(B)$. Consequently, we have

$$r_0(B) \leq r_0(\mathrm{Hom}(B_\tau, B)) = r_0(R) \leq r_0(B).$$

Now let $0 \neq U$ be a pure fully invariant subgroup of B and select $0 \neq u \in U$. Then $Ru \leq U$ and $Ru \cong R$ as an R-module. Hence $r_0(B) = r_0(R) \leq r_0(U)$. So U is a pure subgroup of B with $r_0(U) = r_0(B)$ and hence $U = B$. But this means that B is irreducible as required. □

Similar arguments as those in Corollary 2.1.5 show

Lemma 2.2.3 *Let $B = B_\tau \otimes R$ be an irreducible indecomposable Murley-group. Then $\mathrm{Ext}(A, B)$ is torsion-free if $\mathrm{Ext}(A, B_\tau)$ is torsion-free.*

Proof: We consider the short exact sequence

$$0 \to B_\tau \to B \to D \to 0,$$

where D is torsion-free divisible since B_τ is a pure subgroup of B. By Lemma 1.2.1 we obtain the exact sequence

$$\ldots \mathrm{Hom}(A, D) \xrightarrow{\delta} \mathrm{Ext}(A, B_\tau) \xrightarrow{\alpha} \mathrm{Ext}(A, B) \to \mathrm{Ext}(A, D) = 0.$$

Since $\mathrm{Ext}(A, B_\tau)$ is torsion-free and $\mathrm{Hom}(A, D)$ is divisible also $\mathrm{Im}(\delta)$ is torsion-free divisible and hence $\mathrm{Ker}(\alpha) = \mathrm{Im}(\delta)$ is a direct summand of $\mathrm{Ext}(A, B_\tau)$. So the sequence

$$0 \to \mathrm{Ker}(\alpha) \to \mathrm{Ext}(A, B_\tau) \to \mathrm{Ext}(A, B) \to 0$$

is split-exact and thus $\mathrm{Ext}(A, B)$ is torsion-free. □

Now we characterize the groups A such that $\mathrm{Ext}(A, B)$ is torsion-free for an irreducible indecomposable Murley-group B.

Theorem 2.2.4 *Let $B = B_\tau \otimes R$ be an irreducible indecomposable Murley-group. Then the following hold:*

i) $\mathrm{Ext}(A, B)$ *is torsion-free if and only if* $\mathrm{Ext}(A \otimes R, B)$ *is torsion-free;*

ii) *If A is an R-module then* $\mathrm{Ext}(A, B)$ *is torsion-free if and only if* $\mathrm{Ext}_R(A, B)$ *is torsion-free.*

Proof: i) The short exact sequence

$$0 \to B_0 \to R \to D \to 0,$$

with D torsion-free divisible yields the exactness of

$$0 \to A \otimes B_0 \to A \otimes R \to A \otimes D \to 0,$$

which again induces the exact sequence

$$0 = \mathrm{Hom}(A \otimes D, B) \to \mathrm{Hom}(A \otimes R, B) \xrightarrow{\alpha} \mathrm{Hom}(A \otimes B_0, B)$$

$$\to \mathrm{Ext}(A \otimes D, B) \to \mathrm{Ext}(A \otimes R, B) \to \mathrm{Ext}(A \otimes B_0, B) \to 0.$$

By [AlGoe, Lemma 2.4] we know that $\mathrm{Hom}_{\mathbb{Z}}(R, B) = \mathrm{Hom}_R(R, B) \cong B$. Thus

$$\mathrm{Hom}(A \otimes R, B) \cong \mathrm{Hom}(A, \mathrm{Hom}(R, B)) \cong \mathrm{Hom}(A, B) \cong \mathrm{Hom}(A \otimes B_0, B)$$

such that α is onto. Since $\mathrm{Ext}(A \otimes D, B)$ is torsion-free divisible the remaining sequence

$$0 \to \mathrm{Ext}(A \otimes D, B) \to \mathrm{Ext}(A \otimes R, B) \to \mathrm{Ext}(A \otimes B_0, B) \to 0$$

is split-exact and the equivalence is obvious.

ii) This can be shown by standard homological arguments since we know that $\mathrm{Hom}_{\mathbb{Z}}(A, B) = \mathrm{Hom}_R(A, B)$ by [AlGoe, Lemma 2.4]. $\square$

We directly conclude the following corollary, which is in direct contrast to the case $OT(B) \neq tp(\mathbb{Q})$.

Corollary 2.2.5 *Let $B = B_\tau \otimes R$ be an irreducible indecomposable Murley-group. Then the following are equivalent:*

i) $\mathrm{Ext}(A, B)$ *is torsion-free;*

ii) $\mathrm{Ext}_R(A \otimes R, B)$ *is torsion-free;*

iii) $(A \otimes R)/D \leq \bigoplus_n B$, *where D denotes the maximal divisible subgroup of $A \otimes R$.*

Proof: The equivalence of i) and ii) is clear by Theorem 2.2.4.
Now assume iii). First let $M = A \otimes R \leq B$. Then we have a short exact sequence

$$0 \to M \to B \to B/M \to 0,$$

which induces the exact sequence

$$0 = \operatorname{Hom}(B/M, B) \to \operatorname{Hom}(B, B) \to \operatorname{Hom}(M, B)$$

$$\xrightarrow{\delta} \operatorname{Ext}(B/M, B) \xrightarrow{\alpha} \operatorname{Ext}(B, B) \to \operatorname{Ext}(M, B) \to 0.$$

Since B is a rank-1 R-module, B/M is torsion and $B/(R+M)$ has finite p-components. Now we consider $(M+R)/M \cong R/(M \cap R)$, which has also finite p-components. Consider the short exact sequence

$$0 \to (M+R)/M \to B/M \to B/(M+R) \to 0,$$

in which both of the outside groups have bounded p-components. Thus the same holds for the group in the center. So $\operatorname{Ext}(B/M, B) \cong \prod_{p \in \mathbb{P}} G_p$, where the G_p are finite. Since $\operatorname{Ext}(B, B)$ is torsion-free as B is an irreducible indecomposable Murley-group, the group $\operatorname{Im}(\delta)$ coincides with the torsion subgroup of $\prod_{p \in \mathbb{P}} G_p$, which is nothing else but $\bigoplus_{p \in \mathbb{P}} G_p$. Hence $\operatorname{Im}(\alpha)$ is a torsion-free divisible group and thus a pure subgroup of $\operatorname{Ext}(B, B)$. Hence $\operatorname{Ext}(M, B)$ is torsion-free, too.

Now, let $M \leq \bigoplus_n B$. In this case we consider the short exact sequence

$$0 \to M \cap B_1 \to M \to M/(M \cap B_1) \to 0,$$

where B_1 denotes the firt component of $\bigoplus_n B$. By Lemma 1.2.1 also the sequence

$$0 \to \operatorname{Hom}(M/(M \cap B_1), B) \to \operatorname{Hom}(M, B) \xrightarrow{\alpha} \operatorname{Hom}(M \cap B_1, B)$$

$$\to \operatorname{Ext}(M/(M \cap B_1), B) \to \operatorname{Ext}(M, B) \to \operatorname{Ext}(M \cap B_1, B) \to 0$$

is exact. By induction hypothesis we may assume that the groups $\operatorname{Ext}(M/(M \cap B_1), B)$ and $\operatorname{Ext}(M \cap B_1, B)$ are torsion-free. Hence $\operatorname{Im}(\alpha)$ is a pure subgroup of the rank-1 R-module $\operatorname{Hom}(M \cap B_1, B)$. We directly conclude that α is onto, since $\operatorname{Im}(\alpha) \neq 0$, because otherwise $M \cap B_1 = 0$ and hence $M \leq \bigoplus_{n-1} B$. But then also $\operatorname{Ext}(M, B)$ is torsion-free and we are done.

It remains to show that i) implies iii). So let $\operatorname{Ext}(A, B)$ be torsion-free. Then also $\operatorname{Ext}(A \otimes R, B)$ is torsion-free by Theorem 2.2.4 and $K_B(A \otimes R) = D$ by Theorem 1.7.9. So $K_B((A \otimes R)/D) = 0$ and hence $(A \otimes R)/D \leq \bigoplus_n B$ for some $n \in \mathbb{N}$. □

The next theorem characterizes the subgroups U of the special Murley-groups B with $\operatorname{Ext}(U, B)$ torsion-free.

Theorem 2.2.6 *Let $B = B_\tau \otimes R$ be an irreducible indecomposable Murley-group of idempotent type. If U is p-pure in B for some prime p with $B \neq pB$, then $\mathrm{Ext}(U, B)$ is torsion-free if and only if U_*/U is finite, i.e. U is quasi-equal to its purification U_*.*

Proof: Since U is p-pure in B, we have $r_p(U) = r_p(B) = 1$ and there is a short exact sequence
$$0 \to \mathbb{Z} \to U \to V \to 0,$$
in which $V = pV$ and $V[p] = 0$, because $\mathbb{Z}$ is p-pure in U. Since B is homogeneous and not p-divisible we have $\mathrm{Hom}(V, B) = 0$. Thus we obtain the exact sequence
$$0 = \mathrm{Hom}(V, B) \to \mathrm{Hom}(U, B) \to \mathrm{Hom}(\mathbb{Z}, B) \cong B$$
so that $\mathrm{Hom}(U, B)$ is a non-zero ideal of $B = R$ since B has idempotent type. Moreover, since every pure rank-1 subgroup of B is isomorphic to B_τ, we have an exact sequence
$$0 \to B_\tau \to U_* \to D \to 0,$$
in which D is torsion-free divisible. Hence the group $\mathrm{Hom}(U_*, B)$ is also isomorphic to a non-zero ideal of B. Finally, consider the short exact sequence
$$0 \to U \to U_* \to T \to 0,$$
in which T is torsion. We obtain
$$0 = \mathrm{Hom}(T, B) \to \mathrm{Hom}(U_*, B) \to \mathrm{Hom}(U, B)$$
$$\xrightarrow{\delta} \mathrm{Ext}(T, B) \xrightarrow{\alpha} \mathrm{Ext}(U_*, B) \to \mathrm{Ext}(U, B) \to 0$$
by Lemma 1.2.1. Since B is a PID, $\mathrm{Hom}(U_*, B) \cong B \cong \mathrm{Hom}(U, B)$ so that $\mathrm{Im}(\delta)$ is finite.

If $\mathrm{Ext}(U, B)$ is torsion-free, then $\mathrm{Im}(\alpha)$ is a pure subgroup of the torsion-free divisible group $\mathrm{Ext}(U_*, B)$ and hence it is itself torsion-free and divisible. But $\mathrm{Im}(\delta)$ is algebraically compact so that $\mathrm{Ext}(T, B) \cong \mathrm{Im}(\delta) \oplus Im(\alpha)$. By [Fu1, Theorem 52.3] the group $\mathrm{Ext}(T, B)$ is reduced, so $Im(\alpha) = 0$ and $\mathrm{Ext}(T, B)$ is finite. In analogy to the arguments in Section 1.1 we conclude that also T must be finite and hence we are done.

Conversely, let U be quasi-equal to its purification U_*. Then $\mathrm{Ext}(U, B) \cong \mathrm{Ext}(U_*, B)$ is torsion-free since U_* is a pure subgroup of B and B is a torsion-free splitter. □

In contrast to the case $OT(B) \neq tp(\mathbb{Q})$, cf. Theorem 2.1.11, a torsion-free group A with $K_B(A) = 0$ does not need to satisfy $\mathrm{Ext}(A, B)$ is torsion-free if B is a torsion-free splitter:

Corollary 2.2.7 *Let B be an irreducible indecomposable Murley-group of idempotent type such that $|\text{supp}(B)| \geq 2$ and $r_0(B) \geq 2$. Then B contains a subgroup U such that $\text{Ext}(U,B)$ is not torsion-free.*

Proof: Trivially, B contains a subgroup $W \cong \mathbb{Z} \oplus \mathbb{Z}$. Select $p \neq q \in \text{supp}(B)$ and consider the group B/W. Now we choose a subgroup $U \leq B$ containig W such that $U/W = (B/W)_q$. Then U is q-pure in B. If $\text{Ext}(U,B)$ were torsion-free, then U was quasi-equal to its purification U_*. But $r_p(U) = 2$ while $r_p(U_*) = r_p(B) = 1$, a contradiction. □

Now we are able to describe some extension properties of Murley-goups. Because of Theorem 2.2.6 we consider the question when extensions of torsion-free groups by torsion-free groups are again torsion-free.

Theorem 2.2.8 *Let $B = B_\tau \otimes R$ be an irreducible indecomposable Murley-group. If $U \neq 0$ is a pure subgroup of B and*

$$0 \to U \xrightarrow{\alpha} G \to H \to 0$$

is a short exact sequence with torsion-free $\text{Ext}(H,B)$ and $K_B(G) = 0$ then also the group $\text{Ext}(G,B)$ is torsion-free. In particular, if $U \cong \bigoplus_k B_0$ we have

$$r_0(\text{Hom}(G,B)) > r_0(\text{Hom}(H,B)) + (k-1) \cdot r_0(B).$$

Proof: Since U is a pure subgroup of B also U is homogeneous of type τ. Now let V be a pure rank-1 subgroup of U. Then

$$r_p(V) = r_p(U) = r_p(B) = 1,$$

which implies that U/V is divisible and $\text{Hom}(V,B) \cong B$. Hence we obtain the induced exact sequence

$$0 = \text{Hom}(U/V, B) \to \text{Hom}(U,B) \to \text{Hom}(V,B) \to \text{Ext}(U/V, B)\ldots$$

and conclude that $r_0(\text{Hom}(U,B)) \leq r_0(B) = r_0(R)$. Moreover, we consider the exact sequence

$$0 \to \text{Hom}(H,B) \to \text{Hom}(G,B) \xrightarrow{\alpha^*} \text{Hom}(U,B)$$

$$\xrightarrow{\delta} \text{Ext}(H,B) \to \text{Ext}(G,B) \to \text{Ext}(U,B) \to 0,$$

where $M = \text{Im}(\alpha^*) \neq 0$ is a pure subgroup of $\text{Hom}(U, B)$ since $\text{Ext}(H, B)$ is torsion-free. We obtain the exact sequence

$$0 = \text{Hom}(B/U, B) \to \text{Hom}(B, B) = R \to \text{Hom}(U, B) \to \text{Ext}(B/U, B) \dots$$

and get that R is a pure subgroup of $\text{Hom}(U, B)$ because $\text{Ext}(B/U, B)$ is torsion-free since B/U is torsion-free divisible. So also $M \cap R$ is a pure subgroup of R. But R is a PID, so $M \cap R = 0$ or $M \cap R = R$.

If $M \cap R = 0$, then

$$r_0(B) \geq r_0(\text{Hom}(U, B)) \geq r_0(M) + r_0(R) > r_0(R)$$

which is a contradiction. Thus $M \cap R = R$, which implies that $R \leq M \leq_* \text{Hom}(U, B)$. If now $M \neq \text{Hom}(U, B)$, then

$$r_0(R) \geq r_0(\text{Hom}(U, B)) > r_0(M) \geq r_0(R)$$

is a contradiction, too. Thus $M = \text{Hom}(U, B)$ and so the sequence

$$0 \to U \xrightarrow{\alpha} G \to H \to 0$$

is B-cobalanced. Hence $\text{Ext}(G, B)$ is torsion-free.

So $\text{Im}(\delta)$ is a pure subgroup of $\text{Ext}(H, B)$ and if $U \cong \bigoplus_k B_0$ then $\text{Hom}(U, B) \cong \bigoplus_k \text{Hom}(B_0, B) = \bigoplus_k B$ is an R-module. Thus the short exact sequence

$$0 \to M \to \bigoplus_k B \to \text{Im}(\delta) \to 0$$

splits by [Al4, Cor. 4.10], which implies that $M = \text{Hom}(U, B)$. Thus we directly conclude that

$$r_0(\text{Hom}(G, B)) = r_0(\text{Hom}(H, B)) + r_0(\text{Hom}(U, B)) = r_0(\text{Hom}(H, B)) + k \cdot r_0(B)$$

$$> r_0(\text{Hom}(H, B)) + (k - 1) \cdot r_0(B).$$

$\square$

To get a direct contrast of the cases $OT(B) \neq tp(\mathbb{Q})$ and $OT(B) = tp(\mathbb{Q})$ we just compare Theorem 2.1.12 with the apparent simple case, that B is a rank-2 group. For example, let B be p-local. The structure of these groups was investigated in Theorem 1.6.3.

Our next theorem characterizes the groups A having the property that $\text{Ext}(A, B)$ is torsion-free:

Proposition 2.2.9 *Let A be a torsion-free group of finite rank. Then $\mathrm{Ext}(A, B)$ is torsion-free if and only if $A \otimes B_0 = D \oplus G$ where D is the maximal divisible subgroup of $A \otimes B_0$ and G is a subgroup of $\bigoplus_n B$ with $\mathrm{Ext}(G, B)$ is torsion-free.*

Proof: First let be $\mathrm{Ext}(A, B) \cong \mathrm{Ext}(A \otimes B_0, B)$ torsion-free. In this case we know $K_B(A \otimes B_0) = D$. So we write $A \otimes B_0 = D \oplus G$, where G is a reduced B_0-module and $K_B(G) = 0$. But this implies that $G \leq \bigoplus_n B$ by [HuWa]. Since $\mathrm{Ext}(A \otimes B_0, B)$ is torsion-free, clearly the group $\mathrm{Ext}(G, B)$ has to be torsion-free as well. The other implication is obvious. □

The next proposition characterizes the subgroups $U \leq A$ having torsion-free $\mathrm{Ext}(U, B)$ whenever $\mathrm{Ext}(A, B)$ is torsion-free.

Proposition 2.2.10 *Let A be a $\mathbb{Q}_p$-module such that $\mathrm{Ext}(A, B)$ is torsion-free and let U be a full rank Q_p-submodule of A. Then the following are equivalent:*

i) *$\mathrm{Ext}(U, B)$ is torsion-free;*

ii) *$\mathrm{Ext}(A/U, B)$ is the pure-injective hull of $M = \mathrm{Hom}(U, B)/\mathrm{Hom}(A, B)$.*

Proof: We first consider the short exact sequence

$$0 \to U \to A \to A/U \to 0.$$

Note that A/U is a torsion divisible group without any q-components since A and U are q-divisible as Q_p-modules. So $A/U \cong \bigoplus_n \mathbb{Z}_{p^\infty}$. By Lemma 1.2.1 this induces the exact sequence

$$0 = \mathrm{Hom}(A/U, B) \to \mathrm{Hom}(A, B) \to \mathrm{Hom}(U, B)$$

$$\xrightarrow{\delta} \mathrm{Ext}(A/U, B) \xrightarrow{\beta} \mathrm{Ext}(A, B) \xrightarrow{\alpha} \mathrm{Ext}(U, B) \to 0.$$

Let D denote the divisible hull of B. Hence we have $\mathrm{Ext}(A/U, B) \cong \mathrm{Hom}(A/U, D/B)$ which is isomorphic to a direct sum of n copies of the p-adic integers J_p and thus a reduced algebraically compact group.

i) $\Rightarrow$ ii) If now $\mathrm{Ext}(U, B)$ is torsion-free then $\mathrm{Ker}(\alpha) = \mathrm{Im}(\beta)$ is a pure subgroup of $\mathrm{Ext}(A, B)$ and hence torsion-free divisible. Furthermore it is

$$\mathrm{Im}(\beta) \cong \mathrm{Ext}(A/U, B)/\mathrm{Ker}(\beta) = \mathrm{Ext}(A/U, B)/\mathrm{Im}(\delta) \cong \mathrm{Ext}(A/U, B)/M.$$

Thus $\text{Ext}(A/U)/M$ is divisible and hence $\text{Ext}(A/U, B)$ is the pure-injective hull of M by [Fu1, Lemma 41.8].

ii) $\Rightarrow$ i) Since $\text{Ker}(\alpha) = \text{Im}(\beta) \cong \text{Ext}(A/U, B)/M$ is divisible, the sequence

$$0 \to \text{Ker}(\alpha) \to \text{Ext}(A, B) \to \text{Ext}(U, B) \to 0$$

is split-exact and therefore $\text{Ext}(U, B)$ is torsion-free. $\qquad\square$

We now investigate the different cases which appear in Theorem 1.6.3. Since in i), ii) and iii) we only deal with the trivial case of decomposable groups, we start with case iv).

Lemma 2.2.11 *Every group B as in Theorem 1.6.3 iv) is a torsion-free splitter.*

Proof: Since $OT(B) = tp(\mathbb{Q})$ we know that $r_p(B) < 2$ for all primes $p \in \mathbb{P}$. Furthermore we have $r_p(B) = r_p(B_0)$ because B_0 is p-divisible if and only if B is p-divisible. Hence we directly conclude that

$$r_p(\text{Ext}(B, B)) = (r_p(B))^2 - r_p(\text{Hom}(B, B)) = (r_p(B))^2 - r_p(B_0) = 0.$$

$\qquad\square$

The next lemma tells us, that there are lots of groups A such that $\text{Ext}(A, B)$ is not torsion-free if B is a group of the same form as above.

Lemma 2.2.12 *Let B be a group as in Theorem 1.6.3 iv). If*

$$(E) \quad 0 \to B \to A \to B \to 0$$

represents a non-zero element of $\text{Ext}(B, B)$ then $\text{Ext}(A, B)$ is not torsion-free.

Proof: By Lemma 1.2.1 the short exact sequence (E) induces the exact sequence

$$0 \to \text{Hom}(B, B) \to \text{Hom}(A, B) \to \text{Hom}(B, B)$$
$$\xrightarrow{\delta} \text{Ext}(B, B) \to \text{Ext}(A, B) \to \text{Ext}(B, B) \to 0.$$

We now assume that $\text{Ext}(A, B)$ is torsion-free. Then $\text{Im}(\delta)$ is a pure subgroup of $\text{Ext}(B, B)$ and hence torsion-free divisible. But $\text{Hom}(B, B)$ is isomorphic to B_0 and there are no epimorphisms from B_0 into a torsion-free divisible group except the trivial one. Hence $\text{Im}(\delta) = 0$ and we obtain the sequence

$$0 \to \text{Hom}(B, B) \to \text{Hom}(A, B) \to \text{Hom}(B, B) \to 0,$$

which is split-exact since $\mathrm{Hom}(B,B) \cong B_0$ and $\mathrm{Ext}(B_0, B_0) = 0$. So the induced sequence

$$0 \to \mathrm{Hom}(B_0, B) \to \mathrm{Hom}(\mathrm{Hom}(A,B), B) \to \mathrm{Hom}(B_0, B) \to 0$$

is also split-exact. Furthermore, we have $\mathrm{Hom}(B_0, B) \cong B$ which gives raise to the commutative diagram

$$\begin{array}{ccccccccc} 0 & \to & \mathrm{Hom}(B_0, B) & \to & \mathrm{Hom}(\mathrm{Hom}(A,B), B) & \to & \mathrm{Hom}(B_0, B) & \to & 0 \\ & & \uparrow & & \uparrow\varphi & & \uparrow & & \\ 0 & \to & B & \to & A & \to & B & \to & 0 \end{array}$$

where the map φ must be an isomorphism by the 5-Lemma. Hence the sequence (E) is also split-exact, a contradiction. $\square$

In the following let B be as in Theorem 1.6.3 iv) additionally satisfying $E = E(B) = \mathbb{Q}_p$. Let all groups be $\mathbb{Q}_p$-modules. The first result provides a large class of groups A, for which $\mathrm{Ext}(A, B)$ is torsion-free:

Proposition 2.2.13 $\mathrm{Ext}(\mathrm{Hom}(A,B), B)$ *is torsion-free for every torsion-free group A of finite rank.*

Proof: We choose a finitely generated free submodule F_1 of A such that A/F_1 is torsion. Since A/F_1 is the direct sum of a divisible and a finite group, there is a submodule F of A containing F_1 with F/F_1 finite and A/F divisible. Clearly, F is free, say $F \cong \bigoplus_n \mathbb{Q}_p$ for some $n < \omega$. Then, $\mathrm{Hom}(F, B) \cong \bigoplus_n B$, and there exists an exact sequence

$$0 = \mathrm{Hom}(A/F, B) \to \mathrm{Hom}(A, B) \to \mathrm{Hom}(F, B) \to \mathrm{Ext}(A/F, B)$$

in which $\mathrm{Ext}(A/F, B)$ is a reduced torsion-free group. Therefore, $\mathrm{Hom}(A,B)$ is a pure subgroup of $\bigoplus_n B$, and hence $\mathrm{Ext}(A, B)$ is torsion-free. $\square$

We directly obtain

Corollary 2.2.14 $\mathrm{Ext}(A, B)$ *is torsion-free for all B-reflexive groups A.*

Proof: Since B-reflexive groups A have the property that $\mathrm{Hom}(A, B) = A$ this is clear. $\square$

Moreover, we have the following nice extension-theorem, which is in direct contrast to Theorem 2.1.12.

Theorem 2.2.15 Let B be as in Theorem 1.6.3 iv) and let

$$0 \to \mathbb{Q}_p \oplus \mathbb{Q}_p \xrightarrow{\alpha} A \xrightarrow{\beta} U \to 0$$

be exact with $\text{Ext}(U, B)$ torsion-free. Then $\text{Ext}(A, B)$ is torsion-free if and only if $K_B(A) = 0$ and $r_0(\text{Hom}(A, B)) > r_0(\text{Hom}(U, B)) + 1$.

Proof: Since $\text{Ext}(U, B)$ is torsion-free we have $K_B(U) = p^\omega U = 0$. Now we consider the induced exact sequence

$$0 \to \text{Hom}(U, B) \xrightarrow{\beta^*} \text{Hom}(A, B) \xrightarrow{\alpha^*} \text{Hom}(\mathbb{Q}_p \oplus \mathbb{Q}_p, B)$$

$$\xrightarrow{\delta} \text{Ext}(U, B) \xrightarrow{\beta^*} \text{Ext}(A, B) \to 0$$

and let $M = \text{Im}(\alpha^*) \leq \text{Hom}(\mathbb{Q}_p \oplus \mathbb{Q}_p, B)$. We observe $\text{Hom}(\mathbb{Q}_p \oplus \mathbb{Q}_p, B) \cong B \oplus B$. Suppose that $\text{Ext}(A, B)$ is torsion-free and assume that $r_0(\text{Hom}(A, B)) \leq r_0(\text{Hom}(U, B)) + 1$. As Goeters showed, $K_B(A) = p^\omega A$. Since A is reduced, the latter group vanishes, and $A \leq \bigoplus_n B$ for some $n \in \mathbb{N}$. In particular, $\text{Hom}(A, B) \neq 0$. Thus $r_0(\text{Hom}(A, B)) = r_0(\text{Hom}(U, B))$ or $r_0(\text{Hom}(A, B)) = r_0(\text{Hom}(U, B)) + 1$. In the former case, $\text{Hom}(A, B) \cong \text{Hom}(U, B)$, we obtain $M = 0$ since M is torsion-free. Therefore every homomorphism $\varphi : A \to B$ satisfies $\varphi(U) = 0$ and thus $U \leq K_B(A) = 0$, a contradiction.
Hence $\text{Hom}(A, B)$ has rank $r_0(\text{Hom}(U, B)) + 1$ and M is a rank-1 subgroup of $B \oplus B$. Since $\text{Ext}(A, B)$ is torsion-free, $\text{Im}(\delta)$ is a pure subgroup of rank 3 of the torsion-free divisible group $\text{Ext}(B, B)$. Hence, $\text{Im}(\delta) \cong \mathbb{Q}^3 = \mathbb{Q} \oplus \mathbb{Q} \oplus \mathbb{Q}$. Moreover, $M \cong \mathbb{Q}_p$ yields the induced sequence

$$0 = \text{Hom}(\mathbb{Q}^3, B) \to \text{Hom}(\mathbb{Q}_p \oplus \mathbb{Q}_p, B) \to \text{Hom}(M, B) \cong B,$$

which is not possible. Therefore, $r_0(\text{Hom}(A, B)) > r_0(\text{Hom}(U, B)) + 1$.
Conversely, assume $K_B(A) = 0$ and $r_0(\text{Hom}(A, B)) > r_0(\text{Hom}(U, B)) + 1$.
Since $r_0(\text{Hom}(\mathbb{Q}_p \oplus \mathbb{Q}_p, B)) = 4$, we have $2 \leq r_0(M) \leq 4$. Observe that M is a pure subgroup of $\text{Hom}(\mathbb{Q}_p \oplus \mathbb{Q}_p, B)$ since $\text{Ext}(U, B)$ is torsion-free. Therefore, $r_0(M) = 4$ yields that α^* is onto, and that δ is an isomorphism. In particular, $\text{Ext}(A, B)$ is torsion-free.
On the other hand, if $r_0(M) = 3$, then $\text{Im}(\delta)$ is isomorphic to a rank-1 torsion-free image of $B \oplus B$, and hence $\text{Im}(\delta) \cong \mathbb{Q}$ by Theorem 1.6.4. But then $\text{Ker}(\beta^*)$ splits, and $\text{Ext}(A, B)$ is isomorphic to a direct summand of the trosion-free group $\text{Ext}(U, B)$.

It remains to consider the case that M has rank 2. By Theorem 1.6.4 part ii), $\text{Ker}(\beta^*) = \text{Im}(\delta) \cong \text{Hom}(\mathbb{Q}_p \oplus \mathbb{Q}_p, B)/M$ is either divisible or isomorphic to B. In the former case, $\text{Ext}(A, B)$ is isomorphic to a direct summand of $\text{Ext}(U, B)$. In the latter case, the embedding $M \subseteq \text{Hom}(\mathbb{Q}_p \oplus \mathbb{Q}_p, B)$ splits by Theorem 1.6.4 part ii). In particular, $M \cong B$ by Azumaya's Theorem.

We obtain the commutative diagram

$$\begin{array}{ccccccccc} 0 & \to & \text{Hom}(M, B) & \to & \text{Hom}(\text{Hom}(A, B), B) & \xrightarrow{\beta^{**}} & \text{Hom}(\text{Hom}(U, B), B) & & \\ & & \uparrow \psi & & \uparrow \psi_A & & \uparrow \psi_B & & \\ 0 & \to & \mathbb{Q}_p \oplus \mathbb{Q}_p & \to & A & \to & U & \to & 0. \end{array}$$

A simple diagram chase shows that β^{**} is onto, and that the induced map ψ is one-to-one since $K_B(A) = 0$ guarantees that ψ_G is one-to-one.

However, it is $\text{Hom}(M, B) \cong \text{Hom}(B, B) = \mathbb{Q}_p$, a contradiction. Thus, the case $\text{Im}(\delta) \cong B$ cannot occur. $\square$

We directly derive

Corollary 2.2.16 *Let $0 \to \mathbb{Q}_p \oplus \mathbb{Q}_p \to A \to B \to 0$ be exact. Then, $\text{Ext}(A, B)$ is torsion-free if and only if $K_B(A) = 0$ and $r_0(\text{Hom}(A, B)) > 2$.*

Proof: Just take $U = B$ in Theorem 2.2.15. $\square$

The next extension-lemma is also of interest in this context.

Lemma 2.2.17 *Let B be as in Theorem 1.6.3 iv). Moreover let $U \leq B$ be a subgroup of A with $K_B(A) = 0$ and $\text{Ext}(A/U, B)$ torsion-free.*
Then $\text{Ext}(A, B)$ is torsion-free provided U is isomorphic to $\mathbb{Q}_p$ or B.

Proof: We consider the short exact sequence

$$0 \to U \xrightarrow{i} A \xrightarrow{\pi} A/U \to 0$$

and obtain the exact sequence

$$0 \to \text{Hom}(A/U, B) \to \text{Hom}(A, B) \xrightarrow{i^*} \text{Hom}(U, B)$$
$$\xrightarrow{\delta} \text{Ext}(A/U, B) \xrightarrow{\pi^*} \text{Ext}(A, B) \xrightarrow{\alpha} \text{Ext}(U, B) \to 0$$

by Lemma 1.2.1. Since $\text{Ext}(A/U, B)$ is torsion-free the group $M = \text{Im}(i^*) = \text{Ker}(\delta)$ is a pure subgroup of $\text{Hom}(U, B)$. Furthermore, it is

$M \neq 0$ because otherwise we would have $U \leq \text{Ker}(\varphi)$ for all homomorphisms $\varphi \in \text{Hom}(A, B)$ and this implies $U \leq K_B(A) = 0$, a contradiction. Thus $r_0(M) > 0$. If $r_0(M) = r_0(\text{Hom}(U, B))$ then $M = \text{Hom}(U, B)$ because of purity. Hence $\text{Im}(\delta) = 0$ and this implies $\text{Ext}(A, B)$ is torsion-free and we are done. So we may assume that $r_0(M) < r_0(\text{Hom}(U, B))$.

We now distinguish the two available cases:

In the first case let $U \cong \mathbb{Q}_p$. Then $\text{Hom}(U, B) \cong B$ is of rank 2 and we assume that $r_0(M) = 1$. But then $\text{Ker}(\pi^*) = \text{Im}(\delta) \cong \text{Hom}(U, B)/M$ is divisible. Since $\text{Ker}(\pi^*)$ is also a subgroup of the torsion-free group $\text{Ext}(A/U, B)$ the group $\text{Im}(\pi^*)$ is torsion-free divisible ant thus the sequence

$$0 \to \text{Im}(\pi^*) \to \text{Ext}(A, B) \to \text{Ext}(U, B) \to 0$$

is split-exact. So we conclude that $\text{Ext}(A, B)$ must be torsion-free.

In the second case let $U \cong B$. Then we have $\text{Hom}(U, B) \cong \mathbb{Q}_p$ which leads to a torsion-free $\text{Ext}(A, B)$ because in this case $M = \text{Hom}(U, B)$, as seen before. $\square$

In the remaining part of this chapter we now consider groups as given by Theorem 1.6.3 v).

As analogon of Theorem 2.2.15 we here have:

Theorem 2.2.18 *Let B be as in Theorem 1.6.3 v) and let*

$$0 \to \mathbb{Q}_p \oplus \mathbb{Q}_p \xrightarrow{\alpha} A \xrightarrow{\beta} U \to 0$$

be exact with $\text{Ext}(U, B)$ torsion-free and $K_B(A) = 0$. Then $\text{Ext}(A, B)$ is torsion-free if and only if $r_0(\text{Hom}(A, B)) > r_0(\text{Hom}(U, B)) + 2$.

Proof: This follows immediately from Theorem 2.2.8. $\square$

Note that this shows that there is a big contrast between the cases iv) and v) in Theorem 1.6.3. Hence the property of being irreducible is important to characterize indecomposable Murley-groups.

We also get an analogon of Lemma 2.2.17:

Lemma 2.2.19 *Let $U \leq B$ be a subgroup of A with $K_B(A) = 0$ such that $\text{Ext}(A/U, B)$ is torsion-free. Then $\text{Ext}(A, B)$ is torsion-free provided U is isomorphic to $\mathbb{Q}_p$ or quasi-isomorphic to B.*

Proof: By the same arguments as in the proof of 2.2.17 we may assume that $r_0(M) < r_0(\text{Hom}(U,B))$.

We distinguish two available cases:

In the first case let $U \cong \mathbb{Q}_p$. Then $\text{Hom}(U,B) \cong B$ is of rank 2 and we assume that $r_0(M) = 1$. But then $\text{Ker}(\pi^*) = \text{Im}(\delta) \cong \text{Hom}(U,B)/M$ is divisible. Since $\text{Ker}(\pi^*)$ is also a subgroup of the torsion-free group $\text{Ext}(A/U,B)$ the group $\text{Im}(\pi^*)$ is torsion-free divisible ant thus the sequence

$$0 \to \text{Im}(\pi^*) \to \text{Ext}(A,B) \to \text{Ext}(U,B) \to 0$$

is split-exact. So we conclude that $\text{Ext}(A,B)$ must be torsion-free.

In the second case let be U quasi-isomorphic to B. Then $\text{Ext}(U,B)$ is torsion-free since $\text{Ext}(B,B)$ is. Thus also $\text{Hom}(U,B)$ is quasi-iomorphic to $\text{Hom}(B,B)$ and so $r_0(\text{Hom}(U,B)) = 2$ which leads to a torsion-free $\text{Ext}(A,B)$ because in this case $M = \text{Hom}(U,B)$, as above. □

We have seen that, even in special cases of the group B the results are significantly more difficult as in the case $OT(B) \neq tp(\mathbb{Q})$. Thus one might expect even worse for the rank-3 or higher case. This is the reason why we restrict our attention from now on to the case of $OT(B) \neq tp(\mathbb{Q})$.

Chapter 3

The General Case

Although we are mainly interested in the question when $\text{Ext}(A, B)$ is torsion-free for torsion-free groups A and B of finite rank, we devote this chapter to the infinite rank case. We start with considering filtrations, chain conditions and such.

3.1 Chain Conditions

In [EkHu] P. Eklof introduced the concept of a new invariant, the so-called Γ-invariant of a group A as a new criterion for freeness. In order to do so he needs the well-known result, that a group $A = \bigcup_{\alpha < \kappa} A_\alpha$ is free if and only if $A_{\alpha+1}/A_\alpha$ is free for all $\alpha < \kappa$. This argument is used to show that all Whitehead-groups are free under the assumption $V = L$. In the proof Eklof considers the set

$$E = \{\alpha < \kappa \mid \text{Ext}(A_{\alpha+1}/A_\alpha, \mathbb{Z}) \neq 0\}$$

and shows that the equivalence-class

$$\widetilde{E} = \{X \subseteq \kappa \mid \exists \text{ cub } C \subseteq \kappa \text{ such that } X \cap C = E \cap C\}$$

is an invariant of the group A, called Γ-invariant $\Gamma(A)$. While it is true in ZFC that $\Gamma(A) = 0$ and $\text{Ext}(A_\alpha, \mathbb{Z}) = 0$ for all $\alpha < \kappa$ imply that $\text{Ext}(A, \mathbb{Z}) = 0$, the converse holds only under the assumption $V = L$.

This chapter establishes that this invariant cannot be extended to torsion-freeness in a natural way.

In Theorem 1.7.6, we showed that, for any pure subgroup A' of a torsion-free group A and any group B, the group $\text{Ext}(A', B)$ is a direct summand of the group $\text{Ext}(A, B)$.

In the proof we considered the exact sequence

$$\ldots \operatorname{Hom}(A,B) \xrightarrow{\varphi} \operatorname{Hom}(A',B) \xrightarrow{\delta} \operatorname{Ext}(A/A',B) \xrightarrow{\alpha} \operatorname{Ext}(A,B) \to \operatorname{Ext}(A',B) \to 0$$

and concluded that

$$\operatorname{Ext}(A,B) \cong \operatorname{Im}(\alpha) \oplus \operatorname{Ext}(A',B).$$

Hence we see, that $\operatorname{Ext}(A,B)$ is torsion-free if and only if $\operatorname{Ext}(A',B)$ and $\operatorname{Im}(\alpha)$ are torsion-free. So, it is necessary to ask when $\operatorname{Im}(\alpha)$ is torsion-free.

Note that

$$\operatorname{Im}(\alpha) \cong \operatorname{Ext}(A/A',B)/\operatorname{Ker}(\alpha)$$

and

$$\operatorname{Ker}(\alpha) = \operatorname{Im}(\delta) \cong \operatorname{Hom}(A',B)/\operatorname{Ker}(\delta).$$

Furthermore, we have $\operatorname{Ker}(\delta) = \operatorname{Im}(\varphi) = \varphi(\operatorname{Hom}(A,B))$. Hence $\operatorname{Ker}(\alpha)$ is isomorphic to the group of homomorphisms $\psi \in \operatorname{Hom}(A',B)$ which do not have an extension to a homomorphism $\overline{\psi} \in \operatorname{Hom}(A,B)$.

So $\operatorname{Im}(\alpha)$ is torsion-free if, for example, $\operatorname{Ext}(A/A',B)$ is torsion-free and $\operatorname{Ker}(\alpha)$ is a pure subgroup of it.

Theorem 3.1.1 *Let $A = \bigcup_{\beta<\kappa} A_\alpha$ be a κ-filtration of A with $A_\beta \leq_* A_{\beta+1} \leq_* A$ for all $\beta < \kappa$. Furthermore, for $\alpha : \operatorname{Ext}(A_{\beta+1}/A_\beta, B) \to \operatorname{Ext}(A_{\beta+1}, B)$, let $\operatorname{Ker}(\alpha)$ be a pure subgroup of $\operatorname{Ext}(A_{\beta+1}/A_\beta, B)$ for all $\beta < \kappa$.*
If $\operatorname{Ext}(A_0, B)$ and $\operatorname{Ext}(A_{\beta+1}/A_\beta, B)$ are torsion-free for all $\beta < \kappa$ then $\operatorname{Ext}(A,B)$ is torsion-free.

Proof: The result follows immediately from the arguments above. □

The next example shows that the condition on $\operatorname{Ker}(\alpha)$ in Theorem 3.1.1 is necessary because otherwise this theorem fails, even in the finite rank case.

Example 3.1.2 Let A be the Corner-group of rank 2. Then A is torsion-free, any subgroup $A_0 < A$ of rank 1 is free and the qoutient-group A/A_0 is divisible. Furthermore, we have $\operatorname{End}(A) = \mathbb{Z}$. By applying $\operatorname{Hom}(-, \mathbb{Z})$ to the short exact sequence

$$0 \to \mathbb{Z} \to A \to \mathbb{Q} \to 0$$

we obtain the exact sequence

$$\mathrm{Hom}(\mathbb{Z},\mathbb{Z}) = \mathbb{Z} \to \mathrm{Ext}(\mathbb{Q},\mathbb{Z}) \xrightarrow{\alpha} \mathrm{Ext}(A,\mathbb{Z}) \to 0 = \mathrm{Ext}(\mathbb{Z},\mathbb{Z}).$$

Since $\mathbb{Q}$ is divisible the group $\mathrm{Ext}(\mathbb{Q},\mathbb{Z})$ is torsion-free and hence of the form $\mathrm{Ext}(\mathbb{Q},\mathbb{Z}) = \bigoplus_{2^{\aleph_0}} \mathbb{Q}$. So $\mathrm{Ker}(\alpha)$ cannot be a pure subgroup of $\mathrm{Ext}(\mathbb{Q},\mathbb{Z})$ and hence $\mathrm{Ext}(A,\mathbb{Z})$ cannot be torsion-free.

The natural idea, that $\mathrm{Ext}(A,B)$ is torsion-free if $\mathrm{Ext}(A_{\alpha+1}/A_\alpha, B)$ is torsion-free for all $\alpha < \kappa$ fails in view of Theorem 2.1.12. So we need a more complex condition to get a torsion-free Γ-invariant. In order to do so we define a special kind of filtration.

Definition 3.1.3 *We say that A has a B-**cobalanced filtration** if $A = \bigcup_{\alpha<\kappa} A_\alpha$ such that A_α is B-cobalanced in $A_{\alpha+1}$ for all $\alpha < \kappa$.*

With the help of this definition we may now strengthen Theorem 3.1.1.

Theorem 3.1.4 *Let A and B be torsion-free and A the union of a smooth ascending chain $\{A_\alpha\}_{\alpha<\kappa}$ of subgroups A_α such that $A_0 = 0$,*

i) $0 \to A_\alpha \to A_{\alpha+1}$ *is B-cobalanced, and*

ii) $\mathrm{Ext}(A_{\alpha+1}/A_\alpha, B)$ *is torsion-free for all $\alpha < \kappa$.*

Then also $\mathrm{Ext}(A,B)$ is torsion-free.

Proof: We have to show that the canonical map $\mathrm{Hom}(A,B) \to \mathrm{Hom}(A,B/pB)$ is onto. Therefore take $f \in \mathrm{Hom}(A,B/pB)$. Suppose that the restriction $f \upharpoonright A_\alpha \in \mathrm{Hom}(A_\alpha, B/pB)$ has a lifting to $f_\alpha \in \mathrm{Hom}(A_\alpha, B)$. Since B is injective with respect to

$$0 \to A_\alpha \to_* A_{\alpha+1} \to A_{\alpha+1}/A_\alpha \to 0,$$

there is an extension $g \in \mathrm{Hom}(A_{\alpha+1}, B)$ with $g \upharpoonright A_\alpha = f_\alpha$. We consider the diagram

$$\begin{array}{ccccc}
\mathrm{Hom}(A_{\alpha+1}/A_\alpha, B) & \xrightarrow{\alpha} & \mathrm{Hom}(A_{\alpha+1}/A_\alpha, B/pB) & \to & 0 \\
\downarrow \tau & \circlearrowleft & \downarrow \psi & & \\
\mathrm{Hom}(A_{\alpha+1}, B) & \xrightarrow{\pi} & \mathrm{Hom}(A_{\alpha+1}, B/pB) & & \\
\downarrow & \circlearrowleft & \downarrow & & \\
\mathrm{Hom}(A_\alpha, B) & \to & \mathrm{Hom}(A_\alpha, B/pB) & \to & 0 \\
\downarrow & \circlearrowleft & \downarrow & & \\
\mathrm{Ext}(A_{\alpha+1}/A_\alpha, B) & \to & \mathrm{Ext}(A_{\alpha+1}/A_\alpha, B/pB) & \to & 0
\end{array}$$

and put
$$\delta := f \upharpoonright A_{\alpha+1} - \pi \circ g \in \text{Hom}(A_{\alpha+1}, B/pB).$$

Clearly, $\delta \upharpoonright A_\alpha = 0$ and hence $\delta \in Im(\psi)$. Since α is surjective there exists $h \in \text{Hom}(A_{\alpha+1}/A_\alpha, B)$ with $\psi(\alpha(h)) = \delta$. We define
$$\lambda := \tau(h) \in Hom(A_{\alpha+1}, B)$$
and have $\pi \circ \lambda = \delta$ and $\lambda \upharpoonright A_\alpha = 0$. Now choose $f_{\alpha+1} = g + \lambda$. This implies
$$f_{\alpha+1} \upharpoonright A_\alpha = f_\alpha$$
and
$$\pi \circ f_{\alpha+1} = \pi \circ g + \delta = f \upharpoonright A_{\alpha+1}.$$
Finally, we put $\bar{f}(a) = f_\alpha(a)$ if $a \in A_\alpha$ and see that $\pi \circ \bar{f} = f$. □

Now we define $\Gamma_{tf}(A)$ as the equivalence class $\tilde{E}$ of the set
$$E = \{\alpha < \kappa \mid A_\alpha \text{ is not } B - \text{cobalanced in } A_{\alpha+1}\}.$$

Note that the underlying relation is really an equivalence relation, cf. before. Then $\Gamma_{tf}(A) = 0$ implies that $\text{Ext}(A, B)$ is torsion-free if all groups $\text{Ext}(A_{\alpha+1}/A_\alpha, B)$ are torsion-free by Theorems 3.1.4.

We will now show that the converse does not hold in V=L, but in case of the existence of supercompact cardinals. Therefore we restrict our considerations on groups A of cardinality $\aleph_1$ and the case $B = \mathbb{Z}$.

Proposition 3.1.5 i) Let A be an $\aleph_1$-free group of cardinality $\aleph_1$ which is not free. Then A does not have a $\mathbb{Z}$-cobalanced $\aleph_1$-filtration.

ii) Let G be a reduced group such that $\text{Ext}(G, \mathbb{Z}) \neq 0$ is torsion-free. If A fits into the short exact sequence
$$0 \to \mathbb{Z} \to A \to G \to 0,$$
then $\text{Ext}(A, \mathbb{Z})$ is torsion-free if and only if the sequence splits.

Proof: i) Suppose that A is the union of a smooth ascending chain $\{A_\alpha\}_{\alpha<\kappa}$ of countable subgroups such that A_α is $\mathbb{Z}$-cobalanced in $A_{\alpha+1}$. Then
$$0 \to \text{Ext}(A_{\alpha+1}/A_\alpha, \mathbb{Z}) \to \text{Ext}(A_{\alpha+1}, \mathbb{Z}) \to \text{Ext}(A_\alpha, \mathbb{Z}) \to 0$$

is a short exact sequence. Since A is $\aleph_1$-free, each A_α is free and thus $\mathrm{Ext}(A_{\alpha+1}/A_\alpha, \mathbb{Z}) = 0$. Furthermore, countable Whitehead-groups are free, so $A_{\alpha+1}/A_\alpha$ is free for all α. But this implies A is free, a contradiction.

ii) Supose that
$$0 \to \mathbb{Z} \to A \to G \to 0$$
is a non-splitting short exact sequence, but that $\mathrm{Ext}(A, \mathbb{Z})$ is torsion-free. We consider the induced exact sequence
$$0 \to \mathrm{Hom}(G, \mathbb{Z}) \to \mathrm{Hom}(A, \mathbb{Z}) \xrightarrow{\alpha} \mathrm{Hom}(\mathbb{Z}, \mathbb{Z})$$
$$\to \mathrm{Ext}(G, \mathbb{Z}) \to \mathrm{Ext}(A, \mathbb{Z}) \to 0.$$
Since $\mathrm{Ext}(G, \mathbb{Z})$ is torsion-free α must be onto. So the sequence
$$0 \to \mathrm{Hom}(G, \mathbb{Z}) \to \mathrm{Hom}(A, \mathbb{Z}) \xrightarrow{\alpha} \mathrm{Hom}(\mathbb{Z}, \mathbb{Z}) \to 0$$
is split-exact as $\mathrm{Hom}(\mathbb{Z}, \mathbb{Z}) \cong \mathbb{Z}$. But this implies that the sequence
$$0 \to \mathbb{Z} \to A \to G \to 0$$
splits, too; a contradiction.

Conversely, if
$$0 \to \mathbb{Z} \to A \to G \to 0$$
is split-exact, then $A \cong \mathbb{Z} \oplus G$ and hence $\mathrm{Ext}(A, \mathbb{Z})$ is trivially torsion-free because $\mathrm{Ext}(G, \mathbb{Z})$ is. $\square$

We directly derive

Corollary 3.1.6 *The following statements are undecidable in ZFC:*

i) *If G has cardinality $\aleph_1$ and $\mathrm{Ext}(G, \mathbb{Z})$ is torsion-free, then G has a $\mathbb{Z}$-balanced filtration of countable subgroups.*

ii) *There exists a reduced group H with $\mathrm{Ext}(H, \mathbb{Z})$ torsion-free and for which one can find an exact sequence*
$$0 \to \mathbb{Z} \to G \to H \to 0$$
with $\mathrm{Ext}(G, \mathbb{Z})$ not torsion-free.

Proof: Assume $V = L$. By [EkMe, XII, Cor. 2.11] there exists a coseparable group G of cardinality $\aleph_1$, which is not free. Since Whitehead-groups are free, $\text{Ext}(G, \mathbb{Z}) \neq 0$. By Proposition 3.1.5, G does not have a $\mathbb{Z}$-balanced filtration and if

$$0 \to \mathbb{Z} \to X \to G \to 0$$

represents a non-zero element of $\text{Ext}(G, \mathbb{Z})$, then $\text{Ext}(X, \mathbb{Z})$ is not torsion-free. On the other hand, we assume that the existence of supercompact cardinals is consistent with ZFC. Then it is also consistent with ZFC that every coseparable group is free, cf. [MeSh]. Thus $\text{Ext}(H, \mathbb{Z})$ torsion-free implies that H is free. Clearly, free goups have $\mathbb{Z}$-cobalanced filtrations and if $\text{Ext}(H, \mathbb{Z})$ is torsion-free, then

$$0 \to \mathbb{Z} \to G \to H \to 0$$

is split-exact. $\square$

As examples we consider

Example 3.1.7 **i)** $[V = L]$ *There exist groups $G_1 \leq G_2$ such that $\text{Ext}(G_i, \mathbb{Z})$ is torsion-free for $i = 1, 2$ and $\text{Ext}(G_2/G_1, \mathbb{Z})$ is also torsion-free, but*

$$0 \to G_1 \to G_2 \to G_2/G_1 \to 0$$

is not $\mathbb{Z}$-balanced.

ii) *There exists a group G such that 0 and G are the only $\mathbb{Z}$-balanced subgroups of G.*

Proof: i) Let G be a non-free group with $0 \neq \text{Ext}(G, \mathbb{Z})$ torsion-free and consider a free resolution

$$0 \to F_1 \to F_2 \to G \to 0.$$

It is obvious, that this short exact sequence cannot be $\mathbb{Z}$-cobalanced.
ii) Just let $G = \mathbb{Q}$. $\square$

Altogether we see that in case of existing supercompact cardinals we can define a torsion-free Γ-invariant such that $\text{Ext}(A, \mathbb{Z})$ is torsion-free if and only if $\Gamma_{tf}(A) = 0$ and $\text{Ext}(A_{\alpha+1}/A_\alpha)$ is torsion-free for all $\alpha < \omega_1$. But this leads to the less interesting case of free groups, as we see in the next theorem.

Theorem 3.1.8 *Let B be a torsion-free group of finite rank and let $E = \text{End}(B)$ be hereditary as well as $r_p(E) = (r_p(B))^2$ for all primes $p \in \mathbb{P}$. Then the following are equivalent for a torsion-free group A of cardinality $\aleph_1$:*

i) *A is B-projective;*

ii) *A is the union of a smooth ascending chain $\{A_\alpha\}_{\alpha<\omega_1}$ of subgroups A_α such that $A_0 = 0$, $A_{\alpha+1}/A_\alpha$ reduced and $\text{Ext}(A_{\alpha+1}/A_\alpha, B)$ torsion-free;*

iii) *A is $\aleph_1$-B-projective and there is a filtration $A = \{A_\alpha\}_{\alpha<\kappa}$ such that the set*

$$E_A = \{\alpha < \omega_1 \mid \exists \tau > \alpha \text{ such that } A_\alpha \text{ is not } B-\text{cobalanced in } A_\tau\}$$

is not stationary in $\aleph_1$.

Proof: By our previous results it remains to show that i) is implied by ii) and by iii).

ii) $\Rightarrow$ i) Since $\text{Ext}(A_{\alpha+1}/A_\alpha, B)$ is torsion-free we know $A_{\alpha+1}/A_\alpha \cong D \oplus H_\alpha$, where D is divisible and $K_B(H_\alpha) = 0$. But $D = 0$ since $A_{\alpha+1}/A_\alpha$ is reduced. Thus $A_{\alpha+1}/A_\alpha \cong H_\alpha \leq \prod B$. But $\text{Hom}(B, S_B(\prod_I B)) \cong \text{Hom}(B, \prod_I B) \cong \prod_I \text{End}(B)$ is locally free. Hence countable subgroups are free. But then countable B-generated subgroups are B-projective, i.e. $A_{\alpha+1}/A_\alpha$ is B-projective. Thus A_α is a direct summand of $A_{\alpha+1}$ and this leads to A is B-projective.

iii) $\Rightarrow$ i) Since

$$E_A = \{\alpha < \omega_1 \mid \exists \tau > \alpha \text{ such that } A_\alpha \text{ is not } B-\text{cobalanced in } A_\tau\}$$

is not stationary in $\aleph_1$ there is a cub $C = \{\sigma_\alpha \mid \alpha < \omega_1\}$ such that $E_A \cap C = \emptyset$. Clearly, we also have $A = \{A_{\sigma_\alpha}\}_{\alpha<\kappa}$ and A_{σ_α} is B-cobalanced in A_τ for all $\tau > \sigma_\alpha$. Hence A_{σ_α} is B-cobalanced in A. Moreover, A_{σ_α} is B-projective. By our standard arguments we conclude that A_{σ_α} has to be a direct summand of $A_{\sigma_{\alpha+1}}$ and hence also $A_{\sigma_{\alpha+1}}/A_{\sigma_\alpha}$ is B-projective. Thus also A is B-projective. □

Hence there is no use in defining a torsion-free Γ-invariant by the concept of cobalanced groups.

3.2 B-coseparable Groups

After we tried to characterize when $\text{Ext}(A, B)$ is torsion-free with the help of some torsion-free Γ-invariant, we now walk on a different path.

Although P.A. Griffith had introduced the notion of coseparabiltity in 1968, it has not received much attention in abelian group theory. Before introducing the generalized notion of coseparable groups, we first have a look at the original definition:

Definition 3.2.1 *A group A is called **coseparable** if it is $\aleph_1$-free and every subgroup B of A with A/B finitely generated contains a direct summand H of A such that A/H is finitely generated. The class of all coseparable groups is denoted by $\mathfrak{Cos}$.*

The next lemma is a summary of results by Chase, Griffith, Hiremath and Hausen concerning coseparable groups.

Lemma 3.2.2 *For any reduced group A the following are equivalent:*

i) $\operatorname{Ext}(A, \mathbb{Z})$ *is torsion-free;*

ii) A *is finitely projective;*

iii) A *is coseparable;*

iv) A *is separable and coseparable.*

Proof: See [EkMe, Theorem 2.13]. □

In view of Propostion 1.5.3, replacing "finitely generated" by "finitely B-generated" in the definition of coseparability would lead to immediate counter-examples. So we call a group G **finitely B-presented** if there exists an exact sequence

$$0 \to U \to B^n \to G \to 0$$

for some $n < \omega$ with $S_B(U) = U$, and define the notion of a B-coseparable group as follows:

Definition 3.2.3 *A B-generated group A is B-**coseparable** if, for any subgroup $U \subseteq A$ with A/U finitely B-presented, there exists a direct summand V of A such that $V \subseteq U$ and A/V is B-projective.*

Note that this definition coincides with the original one in case $B = \mathbb{Z}$.

Before we turn to the first main theorem of this section, we need the following technical result.

Proposition 3.2.4 *Let B be a finitely faithful S-group such that $E(B)$ is hereditary.*

  i) *If G is finitely B-presented, then $B = P \oplus T$ where P is B-projective and T is bounded. In particular, G is B-solvable.*

  ii) *A B-generated group A with $K_B(A) = 0$ and $\operatorname{Ext}(A, B)$ torsion-free is locally B-projective.*

Proof: i) Consider an exact sequence $0 \to U \to B^n \to G \to 0$ with $S_B(U) = U$. We obtain the induced commutative diagram

$$\begin{array}{ccccccccc} 0 & \longrightarrow & T_B H_B(U) & \longrightarrow & T_B H_B(B^n) & \longrightarrow & T_B(M) & \longrightarrow & 0 \\ & & \wr \downarrow \theta_U & & \wr \downarrow \theta_{B^n} & & \downarrow & & \\ 0 & \longrightarrow & U & \longrightarrow & B^n & \longrightarrow & T_B(M) & \longrightarrow & 0 \end{array}$$

for some finitely generated submodule M of $H_B(G)$. Then $T_B(M) \cong G$. Moreover, $M = P \oplus T$ where P is projective and T is bounded since E is right hereditary. Thus G has the desired form. Since B is a finitely faithful S-group, bounded B-generated groups are B-solvable.

ii) Since $K_B(A) = 0$ there is a B-cobalanced sequence $0 \to A \to \prod_I B$. Let A_* be the purification of A in $\prod_I B$ and assume $A_* \neq A$. Since B is a faithfully flat S-group, $S_B(\prod_I B)$ is a pure subgroup of $\prod_I B$. Moreover, $S_B(\prod_I B)$ is B-solvable by [Al1] since E is right and left Noetherian by [HuWa2]. Therefore $A_* \subseteq S_B(\prod_I B)$. By Lemma 1.5.5, A_* is B-generated. Observe that A is B-cobalanced in A_*.
The short exact sequence

$$0 \to A \xrightarrow{\mu} A_* \to A_*/A \to 0$$

induces the exact sequence

$$0 = \operatorname{Hom}(A_*/A, B) \to \operatorname{Hom}(A_*, B) \xrightarrow{\mu^*} \operatorname{Hom}(A, B)$$

$$\to \operatorname{Ext}(A_*/A, B) \to \operatorname{Ext}(A_*, B) \to \operatorname{Ext}(A, B) \to 0.$$

Because μ^* is onto $\operatorname{Ext}(A, B) \cong \operatorname{Ext}(A_*, B)/\operatorname{Ext}(A_*/A, B)$. Since $\operatorname{Ext}(A, B)$ is torsion-free, $\operatorname{Ext}(A_*/A, B)$ is divisible as a pure subgroup of the divisible group $\operatorname{Ext}(A_*, B)$. But since A_*/A is torsion, $\operatorname{Ext}(A_*/A, B)$ is also reduced by [Fu1, Theorem 52.3]. However, this is only possible if $\operatorname{Ext}(A_*/A, B) = 0$. We now show that this is not possible unless $A = A_*$.
If $x \in A_*$ satisfies $px \in A$ for a prime p with $B = pB$, then select an element $a \in A$

with $pa = px$. This is possible since A is B-generated. Thus $x \in A$ and $B \neq pB$ whenever $(A_*/A)_p \neq 0$. On the other hand, if p is a prime with $B \neq pB$ then the divisible hull D of B has the property that $(D/B)_p \neq 0$. Otherwise, for $b \in B$, there would exist $d \in D$ such that $pd = b$. Thus $d + B \in (D/B)[p] = 0$ and B is p-divisible. In particular, $(D/B)_p \neq 0$ for all $p \in \mathbb{P}$ such that $(A_*/A)_p \neq 0$. Since D/B is divisible, $\mathrm{Hom}(A_*/A, D/B) \neq 0$ unless $A_* = A$. However, we have an exact sequence

$$0 = \mathrm{Hom}(A_*/A, D) \to \mathrm{Hom}(A_*/A, D/B) \to \mathrm{Ext}(A_*/A, B) = 0.$$

Consequently, $A = A_*$ and so A is a pure subgroup of $S_B(\prod_I B)$.

Since B is flat as a left E-module, B-generated submodules of B-solvable modules are B-solvable by [Al1]. Moreover, $H_B(A)$ is a pure submodule of $H_B(\prod_I B) \cong E^I$. Because $\mathbb{Q}E$ is semi-simple Artinian, $H_B(A)$ is an S-closed submodule of E^I. Since $E(B)$ is a left Noetherian ring by [Ar] and [HuWa2], E^I is a locally projective right E-module by [Al2]. However, once we have shown that S-closed submodules of locally projective $E(B)$-modules are locally projective, then $A \cong T_B H_B(A)$ is locally A-projective.

Let U be an S-closed submodule of a locally projective right E-module M. If X is a finitely generated submodule of U, then there is a finitely generated projective direct summand P of M containing X. Let W be the S-closure of X in P. Since U is S-closed in M, $W \subset B$. Since P/W is a finitely generated non-singular module, and E is a semi-prime hereditary Noetherian ring [HuWa2], P/W is projective [St]. Thus W is a finitely generated projective direct summand of M. Since $W \leq U$ it is also a direct summand of $U \leq M$. $\square$

Now we are able to prove

Theorem 3.2.5 *Let B be a finitely faithful S-group such that E is hereditary. Then the following are equivalent for a torsion-free reduced group A:*

i) *A is B-generated and $\mathrm{Ext}(A, B)$ is torsion-free;*

ii) *A is B-generated and projective with respect to all short exact sequences $0 \to U \to \bigoplus_n B \to (\bigoplus_n B)/U \to 0$ with $S_A(U) = U$;*

iii) *A is B-coseparable;*

iv) *A is B-coseparable and locally B-projective.*

Proof: i) $\Rightarrow$ ii) Consider a short exact sequence

$$0 \to U \to \bigoplus_n B \to (\bigoplus_n B)/U \to 0 \quad (E)$$

with $S_A(U) = U$. If $V = U_*$ is the purification of U in $\bigoplus_n B$ then V is B-generated by Lemma 1.5.5. By [Al1], V is a direct summand of $\bigoplus_n B$. Let

$$\pi : (\bigoplus_n B)/U \to (\bigoplus_n B)/V$$

be the projection.

Let $\phi \in \text{Hom}(A, (\bigoplus_n B)/U)$ and consider the diagram

$$\begin{array}{ccccccccc}
0 & \longrightarrow & U & \longrightarrow & \bigoplus_n B & \xrightarrow{\beta} & (\bigoplus_n B)/U & \longrightarrow & 0 \\
& & \downarrow^{\iota} & & \downarrow^{1_{B^n}} & & \downarrow^{\pi} & & \\
0 & \longrightarrow & V & \longrightarrow & \bigoplus_n B & \xrightarrow{\beta_1} & (\bigoplus_n B)/V & \longrightarrow & 0.
\end{array}$$

Since V is a direct summand of $\bigoplus_n B$, there is $\delta_1 \in \text{Hom}((\bigoplus_n B)/V, \bigoplus_n B)$ with $\beta_1 \delta_1 = id_{(\bigoplus_n B)/V}$. If $\psi \in \text{Hom}(A, \bigoplus_n B)$, then

$$\pi(\phi - \beta\psi) = \pi\phi - \pi\beta\delta_1\pi\phi = \pi\phi - \pi\beta\psi = \pi\phi - \beta_1\delta_1\pi\phi = 0.$$

Thus $\text{Im}(\phi - \beta\psi) \subseteq V/U$.

On the other hand, the sequence $0 \to U \to V \xrightarrow{\beta|V} V/U \to 0$ induces

$$\text{Hom}(A, V) \xrightarrow{(\beta|V)_*} \text{Hom}(A, V/U) \to \text{Ext}(A, U).$$

Since E is hereditary and $S_B(U) = U$, we obtain that U is finitely B-projective. Therefore, $\text{Ext}(A, U)$ is torsion-free as a direct summand of $\bigoplus_\ell \text{Ext}(A, B)$ for some $\ell < \omega$. However, we also have that V/U is bounded: To see this, consider the induced sequence $0 \to H_B(U) \to H_B(V) \to M \to 0$. It yields the commutative diagram

$$\begin{array}{ccccccccc}
0 & \longrightarrow & T_B H_B(U) & \longrightarrow & T_B H_B(V) & \longrightarrow & T_B(M) & \longrightarrow & 0 \\
& & \wr \downarrow \theta_U & & \wr \downarrow \theta_V & & \downarrow & & \\
0 & \longrightarrow & U & \longrightarrow & V & \longrightarrow & V/U & \longrightarrow & 0.
\end{array}$$

Therefore $T_B(M) \cong V/U$ is torsion and $T_B(M/tM) = 0$. Since B is faithfully flat $M = tM$. However, M is a finitely generated E-module and so its additive group is bounded. But then $V/U \cong T_B(M)$ is bounded, too. Therefore $\text{Hom}(A, V/U)$ is also bounded. Now, this is only possible if $(\beta|V)_*$ is onto. Hence there exists

68

$\psi_1 \in \mathrm{Hom}(A, V)$ such that $(\beta|V)_*\psi_1 = \phi - \beta\psi$. Consequently, $\phi = \beta(\psi_1 + \psi)$ and thus A is projective with respect to (E), as required.

ii) $\Rightarrow$ iii) Consider a subgroup U of A such that A/U is finitely B-presented and let $\pi : A \to A/U$ be the canonical projection. Since A/U is finitely B-presented, there is an epimorphism $\phi : \bigoplus_n B \to A/U$ such that $\mathrm{Ker}(\phi)$ is B-generated. By ii), there exists $\psi \in \mathrm{Hom}(A, \bigoplus_n B)$ such that $\phi\psi = \pi$. Then $\mathrm{Ker}(\psi) \subseteq U$. Furthermore, the sequence $0 \to \mathrm{Ker}(\psi) \to A \to \mathrm{Im}(\psi) \to 0$ splits: To see this, observe that $\mathrm{Im}(\psi)$ is B-projective as a B-generated subgroup of $\bigoplus_n B$ because E is hereditary. Thus $\mathrm{Ker}(\psi)$ is a direct summand of A such that $A/\mathrm{Ker}(\psi) \cong \mathrm{Im}(\psi)$ is B-projective by [Al1]. Consequently, A is B-coseparable.

iii) $\Rightarrow$ i) Consider the short exact sequence $0 \to B \xrightarrow{p} B \xrightarrow{\pi} B/pB \to 0$. It induces $\mathrm{Hom}(A, B) \xrightarrow{\pi_*} \mathrm{Hom}(A, B/pB) \to \mathrm{Ext}(A, B) \xrightarrow{(\cdot p)_*} \mathrm{Ext}(A, B)$. Note that $\mathrm{Ext}(A, B)$ is torsion-free if and only if the map π_* is onto. To see this, choose $\varphi \in \mathrm{Hom}(A, B/pB)$. Then $A/\mathrm{Ker}(\varphi)$ is isomorphic to a subgroup of $B/pB \cong \bigoplus_{r_p(B)} \mathbb{Z}/p\mathbb{Z}$. Since B is a finitely faithful S-group, B/pB is B-solvable and the same holds for $A/\mathrm{Ker}(\varphi)$. Because the latter group is finite, there is an exact sequence $0 \to U \to B^\ell \to A/\mathrm{Ker}(\varphi) \to 0$ for some $\ell < \omega$. In view of the fact that $A/\mathrm{Ker}(\varphi)$ is B-solvable and because B is flat as a E-module, U is A-generated. Consequently, $A/\mathrm{Ker}(\varphi)$ is finitely A-presented and there exists $D \subseteq \mathrm{Ker}(\varphi)$ such that $A = D \oplus P$ for some B-projective group P.

Observe that $\mathrm{Ext}(B, B)$ is torsion-free since B is a finitely faithful S-group. Consider the induced sequence $\mathrm{Hom}(P, B) \to \mathrm{Hom}(P, B/pB) \to \mathrm{Ext}(P, B)$ in which the Ext-group is torsion-free. Then the first map has to be onto and there exists $\psi \in \mathrm{Hom}(A, B)$ with $\pi\psi = \varphi$, i.e. π_* is onto.

Since iv) $\Rightarrow$ iii) is trivial, it remains to show that a B-coseparable group A is locally B-projective. Let $0 \neq a \in A$. Since A is reduced there exists a non-zero integer n with $a \notin nA$. Because $A/nA \cong \bigoplus_I \mathbb{Z}/n\mathbb{Z}$ for some index set I, there exist subgroups U and V of A containing nA such that $a \in U$, U/nA finite, and $A/nA = U/nA \oplus V/nA$.

Since the B-generated group $A/V = (U + V)/V \cong U/(U \cap V)$ is also finite as an image of U/nA, there is a short exact sequence $0 \to W \to \bigoplus_n B \xrightarrow{\pi} A/V \to 0$. Moreover, A/V is B-solvable since all bounded groups are B-solvable since B is a finitely faithful S-group. As before $S_B(W) = W$. Since we have already established the equivalence of i), ii) and iii), we obtain that A is projective with respect to the last sequence.

Because $a \notin V$, the canonical epimorphism $\varphi : A \to A/V$ satisfies $\varphi(a) \neq 0$. By the last paragraph, φ lifts to a map $\psi : A \to B$ such that $\varphi = \pi\psi$. In particular, $\psi(a) \neq 0$. Therefore, $R_B(A) = 0$. By what has been already shown, $\text{Ext}(A, B)$ is torsion-free. Thus A must be locally B-projective by Proposition 3.2.4. □

Recall, that a group B is a Murley group if $r_p(B) \leq 1$ for all primes $p \in \mathbb{P}$. Every indecomposable Murley group has a PID as an endomorphism ring. Therefore we obtain

Corollary 3.2.6 *Let B be an indecomposable Murley group. The following are equivalent for a reduced torsion-free group A:*

i) *A is B-generated and $\text{Ext}(A, B)$ is torsion-free;*

ii) *A is B-generated and projective with respect to all short exact sequences*
$$0 \to U \to \bigoplus_n B \to (\bigoplus_n B)/U \to 0 \text{ with } S_A(U) = U;$$

iii) *A is B-coseparable;*

iv) *A is B-coseparable and locally B-projective.*

Proof: Follows from 3.2.5 sine a Murley group satisfies the required conditions on the group B in the assumption. □

Another useful result is

Theorem 3.2.7 *The following are equivalent for a torsion-free group A of finite rank:*

i) *A is projective with respect to all short exact sequences*
$$0 \to U \to \oplus_n A \to G \to 0$$
with $S_A(U) = U$;

ii) *A is a finitely faithful S-group and E is hereditary.*

Proof: i) $\Rightarrow$ ii) The short exact sequence
$$0 \to A \xrightarrow{p} A \to A/pA \to 0$$
induces the exact sequence
$$0 \to \text{Hom}(A, A) \xrightarrow{(\cdot p)_*} \text{Hom}(A, A) \to \text{Hom}(A, A/pA) \to 0$$

because of i). Since
$$\mathrm{Hom}(A, A/pA) \cong \mathrm{Hom}(A/pA, A/pA) \cong \mathrm{Mat}_{r_p(A)}(\mathbb{Z}/p\mathbb{Z})$$
we immediately obtain $(r_p(A))^2 = r_p(E)$.

To show that E is hereditary, let I be an ideal of E with $I^2 = 0$. Since $(I+pE)/pE$ is a nilpotent ideal of the simple ring E/pE, we have $I \subseteq pE$. Then $N(E) \subseteq pE$ and $pN(E) = N(E) \cap pE = N(E)$ yields $N(E) = 0$ because $N(E)$ is pure in E by [Ar]. Thus $\mathbb{Q}E(A)$ is semi-simple Artinian.

Now consider a finitely generated right module M and a short exact sequence
$$0 \to U \xrightarrow{\alpha} \bigoplus_n E \xrightarrow{\beta} M \to 0.$$
It induces the exact sequence
$$T_A(U) \xrightarrow{T_A(\alpha)} T_A(\bigoplus_n E) \xrightarrow{T_A(\beta)} T_A(M) \to 0.$$
Since $K = \mathrm{Im}(T_A(\alpha))$ is A-generated, the top-row of the diagram

$$\begin{array}{ccccccccc}
0 & \longrightarrow & H_A(K) & \longrightarrow & H_A T_A(\oplus_n E) & \longrightarrow & H_A T_A(M) & \longrightarrow & 0 \\
& & & & \uparrow \phi_{\oplus_n E} & & \uparrow \phi_M & & \\
& & & & \oplus_n E & \xrightarrow{\beta} & M & \longrightarrow & 0
\end{array}$$

is exact. Thus ϕ_M is onto since $\phi_{\oplus_n E}$ is an isomorphism. Furthermore, we now additionally assume that M^+ is torsion-free. Then $M \subseteq \bigoplus_k E$ for some $k < \omega$ since $\mathbb{Q}E$ is semi-simple Artinian ([G] or [St]). We obtain the diagram

$$\begin{array}{ccccc}
0 & \longrightarrow & \mathrm{Hom}(A, T_A(M)) & \longrightarrow & \mathrm{Hom}(A, T_A(\oplus_k E)) \\
& & \uparrow \phi_M & & \uparrow \phi_{\oplus_k E} \\
0 & \longrightarrow & M & \longrightarrow & \oplus_k E.
\end{array}$$

Therefore ϕ_M is an isomorphism.

To see that E is hereditary, consider a right ideal I of E and choose another right ideal J such that $I \oplus J$ is essential in E. Since $\mathbb{Q}E$ is semi-simple Artinian, we have $nE \subseteq I \oplus J$ for some non-zero integer n. Then $T_A(I \oplus J)$ is quasi-isomorphic to $T_A(E) \cong A$. Suppose we have already shown that all groups quasi-isomorphic to A are A-projective. Then $T_A(I)$ is A-projective and $I \cong \mathrm{Hom}(A, T_A(I))$ is projective.

Let be B quasi-isomorphic to A. Without loss of generality, we may assume that B is a subgroup of A with A/B finite. Then $r_p(B) = r_p(A)$ and $r_p(E(B)) = r_p(E(A))$. Thus B is a finitely faithful S-group, too. Suppose that L is a maximal right ideal of $E(B)$ with $LB = B$. If L is not essential then $E(B) = L \oplus S$ for some simple module S. Since S^+ is either torsion or divisible, this is not possible. So, L must be essential and $E(B)/L$ is finite since $\mathbb{Q}E(B) \cong \mathbb{Q}E(A)$ is semi-simple Artinian. Therefore B is faithful.

Since A is a finitely faithful S-group, B is finitely A-generated. Thus we can find an exact sequence $\bigoplus_k A \to B \to 0$. Observe that $S_B(A) = A$ since A and B are quasi-isomorphic and B being a S-group guarantees that all bounded B-generated groups are B-solvable. Since B is faithful as an $E(B)$-module, the last sequence splits by Baer's Lemma. Thus B is A-projective.

$ii) \Rightarrow i)$: Since $(r_p(A))^2 = r_p(E)$, the group $\text{Ext}(A, A)$ is torsion-free. Now apply Theorem 3.2.5. $\square$

We combine the above results in the next

Corollary 3.2.8 *The following are equivalent for a torsion-free group B of finite rank:*

i) *B is a finitely faithful S-group and E is hereditary;*

ii) *The following conditions are equivalent for a torsion-free reduced group A:*

    a) *A is B-generated and $\text{Ext}(A, B)$ is torsion-free;*

    b) *A is B-generated and projective with respect to all short exact sequences $0 \to U \to \bigoplus_n B \to (\bigoplus_n B)/U \to 0$ with $S_A(U) = U$;*

    c) *A is a B-coseparable group;*

    d) *A is B-coseparable and locally B-projective.*

Proof: Since the direction i) $\Rightarrow$ ii) was proved in Theorem 3.2.5, we only have to prove the converse. But this is a direct consequence of Theorem 3.2.7 since B is obviously B-coseparable. $\square$

In order to formulate the next result we define a non-singular R-module M to be coseparable if, for every submodule U of M with M/U finitely generated, there exists a direct summand V of M such that $V \subseteq U$ and M/V is finitely generated and projective.

Theorem 3.2.9 *Let B be a finitely faithful S-group with E hereditary and $r_p(E) = (r_p(B))^2$ for all primes $p \in \mathbb{P}$. Then the following are equivalent for A:*

i) If A is a locally B-projective group which is B-coseparable, then $H_B(A)$ is a coseparable right E-module;

ii) If M is a locally projective right E-module which is coseparable, then $T_B(M)$ is B-coseparable.

Proof: i) $\Rightarrow$ ii) By Theorem 3.2.5, $\text{Ext}(A, B)$ is torsion-free. Since every summand H of A is B-generated and also has the property that $\text{Ext}(H, B)$ is torsion-free, we obtain that H is B-coseparable and locally B-projective by again applying Theorem 3.2.5.

Now, let M be a submodule of $H_B(A)$ such that $X = H_B(A)/M$ is finitely generated. Then the torsion subgroup tX is a submodule of X such that X/tX is a finitely generated non-singular module. By [Al3], X/tX is projective. Select a submodule N of $H_B(A)$ containing M such that $N/M = tX$. Then $H_B(A)/N$ is a finitely generated projective module and $H_B(A) = N \oplus P_1$, where P_1 is projective and $N/M \cong tX$ is finitely generated. Since the additive group of N/M is torsion, N/M is bounded. Therefore $T_B(M)$ is a subgroup of $T_B(N)$ such that $T_B(N/M)$ is finitely B-generated and bounded. However, this means that $T_B(N)/T_B(M)$ is finite since B has finite rank. Because B is a finitely faithful S-group, $T_B(N)/T_B(M)$ is B-solvable, and hence finitely B-presented. Since $T_B(N)$ is a direct summand of $A \cong T_B H_B(A)$ as A is locally B-projective, the group $\theta_A(T_B(N))$ is B-coseparable. So we can find a direct summand U of $\theta_A(T_B(N))$ such that $U \subseteq \theta_A(T_B(M))$ and $\theta_A(T_B(N)) = U \oplus P_2$, where P_2 is finitely B-projective. But $T_B H_B(A) = T_B(N) \oplus T_B(P_1)$ so that $A = U \oplus P$ where P is finitely B-projective. Thus $H_B(A) = H_B(U) \oplus H_P(P)$ with $H_B(P)$ finitely generated and projective.

It remains to show $H_B(U) \subseteq M$. For this, let $\varphi \in H_B(U)$. For every $b \in B$, we have $\varphi(b) \in U = \theta_A(T_B(M))$. Thus there are $\beta_1, \ldots, \beta_n \in M$ and $b_1, \ldots, b_n \in B$ with $\varphi(b) = \theta_A(\Sigma_{i=1}^n \beta_i \otimes b_i) = \Sigma_{i=1}^n \beta_i(b_i) \in MB$, $H_B(U)B \subseteq MB$. Therefore $[M + H_B(U)]B = MB$. Since A is B-solvable, $M + H_B(U) = M$ by Lemma 1.5.4.

ii) $\Rightarrow$ i) Let M be a locally projective right E-module which is coseparable. We consider a subgroup U of $T_B(M)$ such that $T_B(M)/U$ is finitely B-presented. Since A/U is B-solvable by Part i) of Proposition 3.2.4, U is B-generated. Also $T_B(M)$ is B-generated and so there exist maps $\varphi_1, \ldots, \varphi_n \in H_B T_B(M)$ such that $T_B(M) = U + (\Sigma_{i=1}^n \varphi_i(B))$. Now, $H_B(U) + \varphi_1(E) + \ldots + \varphi_n(E) = N$ is a submodule of $H_B T_B(M)$ with $NB = H_B T_B(M)B$. By Lemma 1.5.4, $N = H_B T_B(M)$. Choose a submodule V of $H_B(U)$ such that $H_B T_B(M) = V \oplus P$ for some finitely generated projective P. This is possible since ϕ_M is an isomorphism as M is locally projective. We apply T_B to get $T_B H_B T_B(M) = T_B(V) \oplus T_B(P)$ and $T_B(M) = \theta_{T_B(M)}(T_B(V)) \oplus \theta_{T_B(M)}(T_B(P))$. Then $\theta_{T_B(M)}(T_B(V)) = VB \subseteq H_B(U)B \subseteq U$, as desired. □

We finish this section with an immediate consequence of the above.

Corollary 3.2.10 *Let B be a finitely faithful S-group. The following are equivalent for a locally B-projective group G:*

i) *G is B-coseparable;*

ii) *$H_B(G)$ is a coseparable right E-module.*

Proof: See our arguments above. □

Chapter 4

An Application of our Theory: Torsion-free Pairs

In this chapter we deal with various classes of groups. Therefore we introduce some notations: Let $\mathfrak{F}$ be the class of all free groups and let $\mathfrak{D}$ be the class of all divisible ones. Furthermore, we denote the p-divisible groups by $\mathfrak{pD}$ and the p-groups by $\mathfrak{P}$. The torsion-free groups are denoted by $\mathfrak{Tf}$ and the torsion-free ones of finite rank by $\mathfrak{Tff}$.

4.1 Introduction

In this chapter, we want to generalize the concept of cotorsion pairs which was introduced by Luigi Salce in 1977, cf. [Sa]. We start with

Definition 4.1.1 *For any classes $\mathcal{A}$ and $\mathcal{B}$ of groups we define*

i) $\mathcal{A}^* := \{X \mid \text{Ext}(A, X) \text{ is torsion-free for all } A \in \mathcal{A}\}$

ii) $^*\mathcal{B} := \{X \mid \text{Ext}(X, B) \text{ is torsion-free for all } B \in \mathcal{B}\}$

To investigate these classes of groups there are several baswic properties to show first:

Lemma 4.1.2 *Let $\mathcal{A}$ and $\mathcal{B}$ be classes of groups. Then the following hold:*

i) $\mathcal{A} \subseteq {}^*(\mathcal{A}^*)$,

ii) $\mathcal{B} \subseteq (^*\mathcal{B})^*$,

iii) $(^*(\mathcal{A}^*))^* = \mathcal{A}^*$,

iv) $^*((^*\mathcal{B})^*) = {}^*\mathcal{B}$.

Proof: i) Let $A \in \mathcal{A}$. Then we have $\mathrm{Ext}(A, B)$ torsion-free for all $B \in \mathcal{A}^*$ and hence $A \in {}^*(\mathcal{A}^*)$.

ii) Now let $B \in \mathcal{B}$. Then $\mathrm{Ext}(A, B)$ is torsion-free for all $A \in {}^*\mathcal{B}$ and thus $B \in ({}^*\mathcal{B})^*$.

iii) First let $X \in (^*(\mathcal{A}^*))^*$. So $\mathrm{Ext}(A, X)$ is torsion-free for all $A \in {}^*(\mathcal{A}^*)$. Since $\mathcal{A} \subseteq {}^*(\mathcal{A}^*)$ by part i) we see that $\mathrm{Ext}(A, X)$ is torsion-free for all $A \in \mathcal{A}$ and hence $X \in \mathcal{A}^*$. The other inclusion is trivial by part ii) if we choose $\mathcal{A}^*$ instead of $\mathcal{B}$.

iv) First let $X \in {}^*((^*\mathcal{B})^*)$. So $\mathrm{Ext}(X, B)$ is torsion-free for all $B \in ({}^*\mathcal{B})^*$. Since $\mathcal{B} \subseteq ({}^*\mathcal{B})^*$ by part ii) we see that $\mathrm{Ext}(X, B)$ is torsion-free for all $B \in \mathcal{B}$ and hence $X \in {}^*\mathcal{B}$. The other inclusion is trivial by part i) if we choose ${}^*\mathcal{B}$ instead of $\mathcal{A}$. $\square$

Now we can define a torsion-free pair similarly to the notion of cotorsion pairs:

Definition 4.1.3 *A pair $(\mathcal{A}, \mathcal{B})$ of classes of groups is called a **torsion-free pair** if the following properties are satisfied:*

i) $\mathrm{Ext}(A, B)$ *is torsion-free for all $A \in \mathcal{A}$ and $B \in \mathcal{B}$,*

ii) *if $\mathrm{Ext}(A, X)$ is torsion-free for all $A \in \mathcal{A}$ then X must be an element of $\mathcal{B}$,*

iii) *if $\mathrm{Ext}(X, B)$ is torsion-free for all $B \in \mathcal{B}$ then X must be an element of $\mathcal{A}$.*

*In this case we call $\mathcal{A}$ a **torsion-free contravariant class** and $\mathcal{B}$ a **torsion-free covariant class**.*

There are two immediate examples of torsion-free pairs:

Example 4.1.4 *For any classes $\mathcal{A}$ and $\mathcal{B}$ of groups, the pairs $(^*(\mathcal{A}^*), \mathcal{A}^*)$ and $({}^*\mathcal{B}, ({}^*\mathcal{B})^*)$ are torsion-free pairs.*

This follows directly by Definition 4.1.1 and Lemma 4.1.2.

We call $(^*(\mathcal{A}^*), \mathcal{A}^*)$ the **torsion-free pair co-generated by** $\mathcal{A}$ and $({}^*\mathcal{B}, ({}^*\mathcal{B})^*)$ the **torsion-free pair generated by** $\mathcal{B}$.

There are lots of nice properties of torsion-free pairs:

Theorem 4.1.5 *Let $(\mathcal{A}, \mathcal{B})$ be a torsion-free pair. Then $\mathcal{A}$ is closed under isomorphic images and direct sums. On the other hand, $\mathcal{B}$ is closed under isomorphic images and direct products.*

Proof: Clear by the properties of the functor Ext. □

Here we see the first difference between the torsion-free pairs and the cotorsion pairs: While cotorsion pairs are in both components closed unter extensions, the torsion-free pairs are not closed under this operation, as was shown in Example 3.1.2.

Like in [Sa] we can define a partial order on the class, $\mathcal{TF}$, of torsion-free pairs by putting
$$(\mathcal{A}, \mathcal{B}) \leq (\mathcal{A}', \mathcal{B}') \quad if \quad \mathcal{B} \subseteq \mathcal{B}' \; or, equivalently \; \mathcal{A}' \subseteq \mathcal{A}.$$
Then $\mathcal{TF}$ becomes a complete lattice by setting
$$\bigwedge_{i \in I}(\mathcal{A}_i, \mathcal{B}_i) = (^*(\bigcap_{i \in I} \mathcal{B}_i), \bigcap_{i \in I} \mathcal{B}_i)$$
and
$$\bigvee_{i \in I}(\mathcal{A}_i, \mathcal{B}_i) = (\bigcap_{i \in I} \mathcal{A}_i, (\bigcap_{i \in I} \mathcal{A}_i)^*)$$
for a family $\{(\mathcal{A}_i, \mathcal{B}_i)\}_{i \in I}$ of torsion-free pairs.

Using the lattice structure of $\mathcal{TF}$ we can prove:

Corollary 4.1.6 *For any $(\mathcal{A}, \mathcal{B}) \in \mathcal{TF}$ we have*
$$(\mathcal{A}, \mathcal{B}) = (^*(\mathcal{A}^*), \mathcal{A}^*) = (^*\mathcal{B}, (^*\mathcal{B})^*).$$

Proof: Note that we have $\mathcal{A} = {}^*(\mathcal{A}^*)$ and $\mathcal{B} = (^*\mathcal{B})^*$ since $(\mathcal{A}, \mathcal{B})$ is an element of $\mathcal{TF}$. So the asserted equality is obvious. □

The result above is also true for cotorsion pairs, as was shown by H. Pat Goeters in [Goe1]. So, it makes sense that we only consider torsion-free pairs (co-)generated by classes of groups.

The next theorem is a modification of [Sa, Prop. 2.5]:

Theorem 4.1.7 *Let $(\mathcal{A}, \mathcal{B})$ be a torsion-free pair. Then the following are equivalent:*

i) $\mathbb{Z}/p\mathbb{Z} \in \mathcal{A}$;

ii) *Every group $B \in \mathcal{B}$ is p-divisible;*

iii) *$\mathcal{A}$ contains all p-groups.*

Proof: First let $\mathbb{Z}/p\mathbb{Z} \in \mathcal{A}$, i.e. assume i). Then $\text{Ext}(\mathbb{Z}/p\mathbb{Z}, B) \cong B/pB$ is torsion-free for all $B \in \mathcal{B}$ and thus $B/pB = 0$ since it is also torsion. So, B must be p-divisible, as required for ii).

Now we assume ii). Since $\text{Ext}(A, B) = 0$ (and so torsion-free) for p-divisible groups B and p-groups A, this immediately implies that $\mathcal{A}$ contains all p-groups; cf. iii). Clearly, iii) implies i). □

In 2006 L. Strüngmann introduced the notation of $(\mathfrak{A}, \mathfrak{B})$-cotorsion pairs for arbitrary classes of groups $\mathfrak{A}$ and $\mathfrak{B}$, cf. [Str]. Our generalization can be read as follows:

Definition 4.1.8 *Let $\mathcal{A} \subseteq \mathfrak{A}$ be a subclass of $\mathfrak{A}$ and $\mathcal{B} \subseteq \mathfrak{B}$ be a subclass of $\mathfrak{B}$. The pair $(\mathcal{A}, \mathcal{B})$ is called a $(\mathfrak{A}, \mathfrak{B})$-**torsion-free pair** if the following conditions are satisfied:*

  i) *$\text{Ext}(A, B)$ is torsion-free for all $A \in \mathcal{A}$ and $B \in \mathcal{B}$,*

  ii) *if $X \in \mathfrak{B}$ and $\text{Ext}(A, X)$ is torsion-free for all $A \in \mathcal{A}$ then X must be an element of $\mathcal{B}$,*

  iii) *if $X \in \mathfrak{A}$ and $\text{Ext}(X, B)$ is torsion-free for all $B \in \mathcal{B}$ then X must be an element of $\mathcal{A}$.*

*In this case we call $\mathcal{A}$ an $\mathfrak{A}$-**torsion-free contravariant class** and $\mathcal{B}$ an $\mathfrak{A}$-**torsion-free covariant class** of the $(\mathfrak{A}, \mathfrak{B})$-torsion-free pair $(\mathcal{A}, \mathcal{B})$.*

We directly see, that any torsion-free pair is an $(\mathfrak{Ab}, \mathfrak{Ab})$-torsion-free pair, where $\mathfrak{Ab}$ denotes the class of all abelian groups. Also, we see that any $(\mathfrak{A}, \mathfrak{B})$-torsion-free pair equals $(\mathfrak{A}, \mathfrak{B})$ if $(\mathfrak{A}, \mathfrak{B})$ is already a torsion-free pair.

In an analog way as seen above also the $(\mathfrak{A}, \mathfrak{B})$-torsion-free pairs form a complete lattice, denoted by $(\mathfrak{A}, \mathfrak{B})$-$\mathcal{TF}$.

The next theorem sheds some light on the relationship between torsion-free pairs and $(\mathfrak{A}, \mathfrak{B})$-torsion-free pairs:

Theorem 4.1.9 *Let $(\mathcal{V}, \mathcal{W})$ be an $(\mathfrak{V}, \mathfrak{W})$-torsion-free pair. If $(\mathcal{A}, \mathcal{B})$ is either the torsion-free pair generated by $\mathcal{W}$ or the torsion-free pair cogenerated by $\mathcal{V}$ then $(\mathcal{V}, \mathcal{W}) = (\mathcal{A} \cap \mathfrak{V}, \mathcal{B} \cap \mathfrak{W})$.*

Proof: Obviously, we have $(\mathcal{V}, \mathcal{W}) \leq (\mathcal{A} \cap \mathfrak{V}, \mathcal{B} \cap \mathfrak{W})$. For the converse inclusion we may assume w.l.o.g. that there is $B \in \mathcal{B} \cap \mathfrak{W}$ but $B \notin \mathcal{W}$ and that $(\mathcal{A}, \mathcal{B})$ is the torsion-free pair cogenerated by $\mathcal{V}$. Then $\text{Ext}(A, B)$ is torsion-free for all $A \in \mathcal{A}$ and hence $\text{Ext}(V, B)$ is torsion-free for all $V \in \mathcal{V}$ since $\mathcal{V} \subseteq \mathcal{A}$. But we have $B \in \mathfrak{W}$ thus it follows that $B \in \mathcal{W}$. This is a contradiction and so we obtain $(\mathcal{V}, \mathcal{W}) = (\mathcal{A} \cap \mathfrak{V}, \mathcal{B} \cap \mathfrak{W})$. □

Now we want to consider special classes of $(\mathfrak{A}, \mathfrak{B})$-torsion-free pairs. At first we survey the $(\mathfrak{B}, \mathfrak{Ab})$- and the $(\mathfrak{Ab}, \mathfrak{B})$-torsion-free pairs, where $\mathfrak{B}$ denotes the class of bounded groups. If A or B is a bounded group then, trivially, $\text{Ext}(A, B)$ must be bounded, too. Hence $\text{Ext}(A, B) = 0$ because it has to be torsion-free. So the $(\mathfrak{B}, \mathfrak{Ab})$- and $(\mathfrak{Ab}, \mathfrak{B})$-torsion-free pairs coincide with the $(\mathfrak{B}, \mathfrak{Ab})$- and $(\mathfrak{T}, \mathfrak{Ab})$-cotorsion pairs, which were characterized by L. Strüngmann. For example we cite the characterization of the $(\mathfrak{T}, \mathfrak{T})$-cotorsion pairs:

Lemma 4.1.10 *Let $(\mathcal{A}, \mathcal{B})$ be a $(\mathfrak{T}, \mathfrak{T})$-cotorsion pair. Then*

$$\mathcal{A} = \bigoplus_{p \in \pi(\mathcal{A})} \mathfrak{T}_p = \bigoplus_{p \notin \pi(\mathcal{B})} \mathfrak{T}_p$$

and

$$\mathcal{B} = \bigoplus_{p \in \pi(\mathcal{B})} \mathfrak{T}_p = \bigoplus_{p \notin \pi(\mathcal{A})} \mathfrak{T}_p,$$

where $\pi(\mathcal{A}) = \{p \in \mathbb{P} \mid \mathbb{Z}/p\mathbb{Z} \in \mathcal{A}\}$. The maximal element of the lattice of all $(\mathfrak{T}, \mathfrak{T})$-cotorsion pairs is given by $(\mathfrak{T}, \mathfrak{D} \cap \mathfrak{T})$ and the minimal one by $(\{0\}, \mathfrak{T})$.

Proof: See [Str, Theorem 2.2]. □

Since here we only consider torsion-free groups, we now restrict our attention to the $(\mathfrak{Tf}, \mathfrak{Tf})$-torsion-free pairs. In fact, we mainly investigate torsion-free pairs generated by torsion-free groups of finite rank.

4.2 The Torsion-free Pairs Generated by Torsion-free Groups of Finite Rank

In this section all groups are assumed to be torsion-free.

Theorem 4.2.1 *Recall, we write $\mathfrak{F}$ for the class of free groups, $\mathfrak{D}$ for the divisible ones and $\mathfrak{Tff}$ for the torsion-free groups of finite rank.*
The following holds in the lattice $(\mathfrak{Tff}, \mathfrak{Tff})$-$\mathcal{TF}$:

i) $\mathbf{1} = ((\mathfrak{F} \oplus \mathfrak{D}) \cap \mathfrak{Tff}, \mathfrak{Tff})$,

ii) $\mathbf{0} = (\mathfrak{Tff}, \mathfrak{D} \cap \mathfrak{Tff})$.

Proof: Clear. □

Note, we directly obtain that, on the one hand, the $(\mathfrak{Tff}, \mathfrak{Tff})$-torsion-free pair cogenerated by $\mathbb{Z}$ is the maximal one and, on the other hand, the $(\mathfrak{Tff}, \mathfrak{Tff})$-torsion-free pair generated by $\mathbb{Q}$ is the minimal one.

As an immediate consequence of our investigations in Chapter 1 we obtain

Lemma 4.2.2 *Let $(\mathcal{A}, \mathcal{B})$ be a $(\mathfrak{Tff}, \mathfrak{Tff})$-torsion-free pair. Then $\mathcal{A}$ is closed under taking pure subgroups and $\mathcal{B}$ is closed under epimorphic images.*

Proof: This follows from Corollary 1.7.8. □

We now describe the singly generated $(\mathfrak{Tff}, \mathfrak{Tff})$-torsion-free pairs. These are $(\mathfrak{Tff}, \mathfrak{Tff})$-torsion-free pairs generated by only one group B, which we may assume to be reduced.

Lemma 4.2.3 *Let B be a group with $OT(B) \neq tp(\mathbb{Q})$. Then we have*

$$^*B = \{A \in \mathfrak{Tff} \mid OT((A \otimes B_0)/D) \leq IT(B)\},$$

where D is the divisible subgroup of $A \otimes B_0$.

Proof: By Theorem 2.1.4, the group $\text{Ext}(A, B)$ is torsion-free if and only if $OT((A \otimes B_0)/D) \leq IT(B)$. Thus

$$^*B = \{A \in \mathfrak{Tff} \mid OT((A \otimes B_0)/D) \leq IT(B)\}.$$

□

P. Goeters has shown in [Goe, Theorem 3.1] that, for a countable group A, a finite rank group B with $OT(B) \neq tp(\mathbb{Q})$, and a rational group R with $tp(R) = IT(B)$, the group $\text{Ext}(A, B)$ is torsion-free if and only if $\text{Ext}(A, R)$ is torsion-free. Due to this result we have the next theorem. Note that our proof is easier.

Theorem 4.2.4 *Let B be a group with $OT(B) \neq tp(\mathbb{Q})$. Then there is a rank-1 group R such that $^*B = {}^*R$.*

Proof: We choose R with $tp(R) = IT(B)$. Then $\mathrm{supp}(B) = \mathrm{supp}(R)$ and hence $B_0 = R_0$. So the assertion follows by Lemma 4.2.3. □

An easy implication is

Corollary 4.2.5 *Let B be a group with $OT(B) \neq tp(\mathbb{Q})$. Then there is a rank-1 group R such that the $(\mathfrak{T}\mathrm{ff}, \mathfrak{T}\mathrm{ff})$-torsion-free pair generated by B is exactly the $(\mathfrak{T}\mathrm{ff}, \mathfrak{T}\mathrm{ff})$-torsion-free pair generated by R.*

Proof: Since $^*B = {}^*R$ if $tp(R) = IT(B)$ we have
$$(^*B, (^*B)^*) = (^*R, (^*R)^*).$$
□

Here we also see that two different finite rank groups B and B' generate the same $(\mathfrak{T}\mathrm{ff}, \mathfrak{T}\mathrm{ff})$-torsion-free pair if their outer types are not the type of $\mathbb{Q}$ and $IT(B) = IT(B')$.

If $(\mathcal{A}, \mathcal{B})$ is a $(\mathfrak{T}\mathrm{ff}, \mathfrak{T}\mathrm{ff})$-torsion-free pair generated or co-generated by a rational group, we call $(\mathcal{A}, \mathcal{B})$ a **rational generated**, respectively, **rational co-generated** $(\mathfrak{T}\mathrm{ff}, \mathfrak{T}\mathrm{ff})$-torsion-free pair. Now we specify the structure of the rational generated $(\mathfrak{T}\mathrm{ff}, \mathfrak{T}\mathrm{ff})$-torsion-free pairs.

In the case of cotorsion pairs one can show that, for rank-1 groups R and S with $tp(R) \leq tp(S)$, we have $\mathrm{Ext}(R, B) = 0$ if $\mathrm{Ext}(S, B) = 0$, cf. [GöShWa, Lemma 1.1]. This fails in the case of torsion-free pairs. For example, we have $tp(\mathbb{Q}_{p,q}) \leq tp(\mathbb{Q}_p)$ and $\mathrm{Ext}(\mathbb{Q}_p, \mathbb{Q}^p)$ is torsion-free but $\mathrm{Ext}(\mathbb{Q}_{p,q}, \mathbb{Q}^p)$ is not torsion-free since
$$r_q(\mathrm{Ext}(\mathbb{Q}_{p,q}, \mathbb{Q}^p)) = r_q(\mathbb{Q}_{p,q}) \cdot r_q(\mathbb{Q}^p) - r_q(\mathrm{Hom}(\mathbb{Q}_{p,q}, \mathbb{Q}^p)) = 1 \cdot 1 - 0 = 1.$$

Instead of this we are able to show

Lemma 4.2.6 *Let R and T be rank-1 groups with $tp(R) \leq tp(T)$. Then we have $^*R \subseteq {}^*T$.*

Proof: First note that the assertion is trivial, if T is divisible. Hence we may assume w.l.o.g. that $tp(T) \neq tp(\mathbb{Q})$. Now let $A \in {}^*R$, i.e. $\mathrm{Ext}(A, R)$ is torsion-free. We have to show that $\mathrm{Ext}(A, T)$ is also torsion-free. In order to do so we write

$T = T' \otimes T_0$, where T_0 denotes, as usual, the nucleus of T and T' is defined by $h_p(T') = h_p(T)$ for $h_p(T) \neq \infty$ and $h_p(T') = 0$ otherwise. If D_R denotes the maximal divisible subgroup of $A \otimes R_0$ then D_R equals the maximal divisible subgroup of $A \otimes (T' \otimes R_0)$ and hence we obviously have

$$OT((A \otimes (T' \otimes R_0))/D_R) = OT((A \otimes R_0)/D_R) \leq tp(R) \leq tp(T' \otimes R_0).$$

So the group $\text{Ext}(A, T' \otimes R_0)$ is torsion-free. Now we consider the short exact sequence

$$0 \to T' \otimes R_0 \to T \to T/(T' \otimes R_0) \to 0.$$

Since $T/(T' \otimes R_0) \cong \bigoplus_{p \in T_0 \setminus R_0} \mathbb{Z}_{p^\infty}$ we obtain the exact sequence

$$\cdots \to \text{Hom}(A, \bigoplus_{p \in T_0 \setminus R_0} \mathbb{Z}_{p^\infty}) \xrightarrow{\alpha} \text{Ext}(A, T' \otimes R_0) \to \text{Ext}(A, T) \to 0.$$

But $\text{Ker}(\alpha)$ is a pure subgroup of the divisible group $\text{Hom}(A, \bigoplus_{p \in T_0 \setminus R_0} \mathbb{Z}_{p^\infty})$ since $\text{Ext}(A, T' \otimes R_0)$ is torsion-free. So $\text{Ker}(\alpha)$ is divisible and hence $\text{Ext}(A, T)$ is a direct summand of $\text{Ext}(A, T' \otimes R_0)$ and thus it must be torsion-free as well. □

This result can be improved:

Lemma 4.2.7 *Let R and T be rank-1 groups with $tp(R) < tp(T)$. Then we have $*(R) \subsetneq *(T)$.*

Proof: First let $tp(T) \neq tp(\mathbb{Q})$. Then $\text{Ext}(T, T)$ is torsion-free by Proposition 2.1.8 and hence $T \in {}^*T$. Since $tp(R) < tp(T) \neq tp(\mathbb{Q})$ there is a prime p such that T, and thus R, are not p-divisible. Furthermore, this assumption means $\text{Hom}(T, R) = 0$. So we see that

$$r_p(\text{Ext}(T, R)) = r_p(T) \cdot r_p(R) - r_p(\text{Hom}(T, R)) = 1 \cdot 1 - 0 = 1$$

and this implicates $T \notin {}^*R$. Thus the asssertion is correct.

Now let $tp(T) = tp(\mathbb{Q})$. This means that T is divisible and hence $\text{Ext}(X, T)$ is torsion-free for any group X. Since $tp(R) < tp(T)$ there is at least one prime p such that R is not p-divisible. If there is exactly one prime p fulfilling this condition, then we have $R \cong \mathbb{Q}_p$. Let X be a finite rank pure subgroup of J_p with $r_0(X) > 1$. Then X is q-divisible for all primes $q \neq p$ and thus a $\mathbb{Q}_p$-module. Furthermore, X must be indecomposable by [Fu1, Theorem 88.1]. Hence we have $\text{Hom}(X, \mathbb{Q}_p) = 0$, because otherwise any homomorphism $\varphi \neq 0$ would be an epimorphism since $\mathbb{Q}_p$ is a rational group. Thus we have a short exact sequence

$$0 \to \mathrm{Ker}(\varphi) \to X \xrightarrow{\varphi} \mathbb{Q}_p \to 0.$$

Since the nucleus of $\mathrm{Ker}(\varphi)$ equals $\mathbb{Q}_p$ we have $\mathrm{Ext}(\mathbb{Q}_p, \mathrm{Ker}(\varphi)) = 0$ and hence the sequence must be splitting-exact. So $\mathrm{Ker}(\varphi)$ is a direct summand of X. Since X is indecomposable and $\varphi \neq 0$ we conclude $\mathrm{Ker}(\varphi) = 0$ and hence φ is an isomorphism. But this means that X is a rank-1 group, a contradiction. So we have

$$r_p(\mathrm{Ext}(X, \mathbb{Q}_p)) = r_p(X) \cdot r_p(\mathbb{Q}_p) = r_p(X) \neq 0$$

and hence $\mathrm{Ext}(X, \mathbb{Q}_p)$ is not torsion-free.

If there are two or more primes p such that R is not p-divisible we let X be a rank-1 group, which is not p-divisible for one of these primes, but q-divisible for another prime q such that R is not q-divisible, because then we have $\mathrm{Hom}(X, R) = 0$ and hence again

$$r_p(\mathrm{Ext}(X, R)) = r_p(X) \cdot r_p(R) - r_p(\mathrm{Hom}(X, R)) = 1 \cdot 1 - 0 = 1.$$

This means that $^*R \subsetneq {}^*T$. □

In [Str, Cor. 2.14] it is shown, that the lattice of all singly generated $(\mathfrak{T}\mathfrak{f}\mathfrak{f}, \mathfrak{T}\mathfrak{f}\mathfrak{f})$-cotorsion pairs is anti-isomorphic to the lattice of all idempotent types. In the context of $(\mathfrak{T}\mathfrak{f}\mathfrak{f}, \mathfrak{T}\mathfrak{f}\mathfrak{f})$-torsion-free pairs this result can be modified as follows:

Theorem 4.2.8 *The lattice of types is anti-isomorphic to the lattice of all rational generated $(\mathfrak{T}\mathfrak{f}\mathfrak{f}, \mathfrak{T}\mathfrak{f}\mathfrak{f})$-torsion-free pairs.*

Proof: Remembering the lattice-structure of the $(\mathfrak{T}\mathfrak{f}\mathfrak{f}, \mathfrak{T}\mathfrak{f}\mathfrak{f})$-torsion-free pairs and the results of the Lemmas 4.2.6 and 4.2.7, we directly see that

$$(^*T, (^*T)^*) \leq (^*R, (^*R)^*)$$

if and only if $tp(R) \leq tp(T)$. So the isomorphism φ is given by putting

$$\varphi(R) = (^*R, (^*R)^*).$$

□

By Theorem 1.7.12 we derive the following

Corollary 4.2.9 *Let A be of finite rank and let A' be a pure subgroup of A. Then $A^* \subseteq (A')^*$.*

Proof: Let $B \in A^*$. Then $\text{Ext}(A, B)$ is torsion-free and hence $\text{Ext}(A', B)$ is torsion-free by Corollary 1.7.8. So we also have $B \in (A')^*$ and thus $A^* \subseteq (A')^*$. □

Analogously we see

Corollary 4.2.10 *Let B be of finite rank with $OT(B) \neq tp(\mathbb{Q})$ and let B' be a pure subgroup of B. Then $^*B \subseteq {}^*(B')$.*

Proof: Let $A \in {}^*B$. Then $\text{Ext}(A, B)$ is torsion-free and hence $\text{Ext}(A, B')$ is torsion-free by Theorem 1.7.12. So we also have $A \in {}^*(B')$ and thus $^*B \subseteq {}^*(B')$. □

This cannot be improved like Lemma 4.2.6 to Lemma 4.2.7 since a group B with $OT(B) \neq tp(\mathbb{Q})$ can have a proper pure subgroup B' with $IT(B') = IT(B)$. For example, we choose $B = \mathbb{Z} \oplus \mathbb{Q}_p$ and $B' = \mathbb{Z}$. Then $\mathbb{Z}$ is a proper pure subgroup of B but we have $IT(\mathbb{Z}) = IT(B)$ and hence $^*B = {}^*\mathbb{Z}$.

In the next step, we consider the $(\mathfrak{T}\text{ff}, \mathfrak{T}\text{ff})$-torsion-free pairs generated by a class of groups. To simplify the notation we put $\mathcal{B} = \{B_i \mid i \in I\}$ for an index set I and restrict our investigations to groups B_i with $OT(B_i) \neq tp(\mathbb{Q})$.

Theorem 4.2.11 *Let $(\mathcal{A}, \mathcal{B})$ be the $(\mathfrak{T}\text{ff}, \mathfrak{T}\text{ff})$-torsion-free pair generated by $\mathcal{B}' = \{B_i \mid i \in I\}$ for an index set I and $OT(B_i) \neq tp(\mathbb{Q})$ for all $i \in I$. Then $(\mathcal{A}, \mathcal{B})$ is generated by a class of rank-1 groups.*

Proof: First note, that we have $A \in {}^*(\mathcal{B}')$ if and only if the group $\text{Ext}(A, B_i)$ is torsion-free for all $i \in I$, which is equivalent to $A \in {}^*(B_i)$ for all $i \in I$. That means $A \in \bigcap_{i \in I} {}^*B_i$. Hence we see that ${}^*(\mathcal{B}') = \bigcap_{i \in I} {}^*B_i$ and therefore it is

$$({}^*(\mathcal{B}'), ({}^*(\mathcal{B}'))^*) = (\bigcap_{i \in I} {}^*(B_i), (\bigcap_{i \in I} {}^*B_i)^*).$$

By Corollary 4.2.5 we know that $^*B_i = {}^*R_i$ for rational groups R_i with $tp(R_i) = IT(B_i)$. This means that $(\mathcal{A}, \mathcal{B})$ is generated by $\mathcal{B}^* = \{R_i \mid i \in I\}$. □

In case of a finite index set I we can prove a stricter version:

Corollary 4.2.12 *Let $(\mathcal{A}, \mathcal{B})$ be the $(\mathfrak{T}\mathit{ff}, \mathfrak{T}\mathit{ff})$-torsion-free pair generated by $\mathcal{B}' = \{B_i \mid i \in I\}$ for a finite index set I, $OT(\bigoplus_{i \in I} B_i) \neq tp(\mathbb{Q})$. Then $(\mathcal{A}, \mathcal{B})$ coincides with the $(\mathfrak{T}\mathit{ff}, \mathfrak{T}\mathit{ff})$-torsion-free pair generated by a rank-1 group R with $tp(R) = \inf\{IT(B_i) \mid i \in I\}$.*

Proof: Let $B = \bigoplus_{i \in I} B_i$. Then we trivially have $\mathcal{A} = {}^*B$ because B has finite rank. Since $OT(\bigoplus_{i \in I} B_i) \neq tp(\mathbb{Q})$ we also have $OT(B_i) \neq tp(\mathbb{Q})$ for all $i \in I$. Hence $(\mathcal{A}, \mathcal{B})$ is generated by $\mathcal{B}^* = \{R_i \mid i \in I\}$, where the R_i are rational groups with $tp(R_i) = IT(B_i)$. Therefore we conclude

$$({}^*(\mathcal{B}'), ({}^*(\mathcal{B}'))^*) = (\bigcap_{i \in I} {}^*R_i, (\bigcap_{i \in I} {}^*R_i)^*)$$

and thus $(\mathcal{A}, \mathcal{B}) = \bigvee_{i \in I}({}^*R_i, ({}^*R_i)^*)$ by the remark before Corollary 4.1.6. Now recall, that the lattice of types is anti-isomorphic to the lattice of all rational generated $(\mathfrak{T}\mathit{ff}, \mathfrak{T}\mathit{ff})$-torsion-free pairs by Theorem 4.2.8. Hence

$$\bigvee_{i \in I}({}^*R_i, ({}^*R_i)^*)$$

must be generated by a rational group R with

$$tp(R) = IT(B) = \inf\{tp(R_i) \mid i \in I\} = \inf\{IT(B_i) \mid i \in I\}.$$

$\square$

The above results fail if $OT(\bigoplus_{i \in I} B_i) = tp(\mathbb{Q})$, even in the case of a finite index set I. For example, consider $\mathcal{B}' = \{\mathbb{Q}_p, \mathbb{Q}^p\}$ and $B = \mathbb{Q}_p \oplus \mathbb{Q}^p$. Then we trivially have $OT(B) = tp(\mathbb{Q})$. Since $\mathrm{Ext}(\mathbb{Q}_p, \mathbb{Q}_p)$ and $\mathrm{Ext}(\mathbb{Q}_p, \mathbb{Q}^p)$ are torsion-free we conclude that $\mathbb{Q}_p \in {}^*B$ but $\inf\{tp(\mathbb{Q}_p), tp(\mathbb{Q}^p)\} = tp(\mathbb{Z})$ and $\mathrm{Ext}(\mathbb{Q}_p, \mathbb{Z})$ is not torsion-free.

So we obtain:

Corollary 4.2.13 *The lattice of all rational generated $(\mathfrak{T}\mathit{ff}, \mathfrak{T}\mathit{ff})$-torsion-free pairs is a proper sublattice of the lattice of all $(\mathfrak{T}\mathit{ff}, \mathfrak{T}\mathit{ff})$-torsion-free pairs.*

Proof: Clear by the remark above. $\square$

This result gives raise to the natural question of the structure of the complete lattice of $(\mathfrak{T}\mathit{ff}, \mathfrak{T}\mathit{ff})$-torsion-free pairs. By Theorem 4.2.11 we may restrict our considerations to $(\mathfrak{T}\mathit{ff}, \mathfrak{T}\mathit{ff})$-torsion-free pairs which are generated by a class of rank-1 groups. The first result is easily established:

Lemma 4.2.14 *Let $\mathcal{B} = \{R_i \mid i \in I\}$ and $\mathcal{B}' = \{R_j \mid j \in J\}$ be classes of rank-1 groups with $\mathcal{B} \subseteq \mathcal{B}'$. Then we have $^*(\mathcal{B}') \subseteq {}^*\mathcal{B}$.*

Proof: Let $A \in {}^*(\mathcal{B}')$. Then $\mathrm{Ext}(A, R_j)$ is torsion-free for any $R_j \in \mathcal{B}'$. Since we have $\mathcal{B} \subseteq \mathcal{B}'$ the group $\mathrm{Ext}(A, R_i)$ must be torsion-free for all $R_i \in \mathcal{B}$ and hence we have $A \in {}^*\mathcal{B}$. □

Furthermore, we need the next technical lemma:

Lemma 4.2.15 *Let A be a reduced torsion-free group of finite rank which is not free. Then there exists a rank-1 group R such that $\mathrm{Ext}(A, R)$ is not torsion-free.*

Proof: Since A is reduced there is a prime $p \in \mathbb{P}$ such that A is not p-divisible. We now distinguish two cases:

i) A is q-divisible for at least one prime $q \in \mathbb{P} \setminus \{p\}$
Then we choose $R = \mathbb{Q}_{p,q}$ because then we have $\mathrm{Hom}(A, R) = 0$ since A is q-divisible but R is not. Thus we have
$$r_p(\mathrm{Ext}(A, R)) = r_p(A) \cdot r_p(\mathbb{Q}_{p,q}) \neq 0$$
since A and $\mathbb{Q}_{p,q}$ are not p-divisible and therefore $\mathrm{Ext}(A, R)$ is not torsion-free.

ii) A is not p-divisible for all primes $p \in \mathbb{P}$
In this case we choose $R = \mathbb{Z}$ because then we have $OT((A \otimes R_0)/D) = OT(A)$ since A is reduced. Moreover, it is $OT(A) \neq tp(\mathbb{Z})$ as A is not free. So we conclude that $OT(A) > tp(\mathbb{Z})$ and hence $\mathrm{Ext}(A, R)$ is not torsion-free.

□

It is now straight forward to prove the next result.

Theorem 4.2.16 *Let $\mathcal{B} = \{R_i \mid i \in \mathbb{N}\}$ be a countable infinite class of rank-1 groups where $tp(R_i) \neq tp(\mathbb{Q})$ for at least one $i \in \mathbb{N}$. Then there exists a rational group R such that $^*(\mathcal{B} \cup \{R\}) \subsetneqq {}^*\mathcal{B}$.*

Proof: Since there is an $i \in I$ with $tp(R_i) \neq tp(\mathbb{Q})$ the $(\mathfrak{T}\mathrm{ff}, \mathfrak{T}\mathrm{ff})$-torsion-free pair generated by $\mathcal{B}$ is not the maximal one. Hence there exists a group $A \in {}^*\mathcal{B}$ which is reduced and not free. By Lemma 4.2.15 we know that there is a rank-1 group R such that $\mathrm{Ext}(A, R)$ is not torsion-free which means that $A \notin {}^*R$ and hence
$$A \notin {}^*R \cap {}^*\mathcal{B} = {}^*(\mathcal{B} \cup \{R\}).$$

Now we consider a descending chain of $(\mathfrak{T}\mathfrak{f}\mathfrak{f}, \mathfrak{T}\mathfrak{f}\mathfrak{f})$-torsion-free pairs $(\mathcal{A}_i, \mathcal{B}_i)$ with $i \in I$ for an index set I. This means we have $(\mathcal{A}_i, \mathcal{B}_i) \geq (\mathcal{A}_j, \mathcal{B}_j)$ for any $i \leq j \in I$. Then, by definition, we obtain an ascending chain $\mathcal{A}_0 \subseteq \mathcal{A}_1 \subseteq \mathcal{A}_2 \subseteq \ldots$. Since $\mathcal{A}_i = {}^*\mathcal{B}_i$ and since there are only groups of finite rank in the class $\mathcal{A}_i$, we conclude that $|\mathcal{A}_i| \leq 2^{\aleph_0}$ for any $i \in I$ (up to isomophism). Hence the ascending chain has length at most $2^{\aleph_0}$ and this means that also the descending chain of the $(\mathcal{A}_i, \mathcal{B}_i)$ has length at most $2^{\aleph_0}$.

Now we turn our attention to the Murley groups of Section 2.3.

Theorem 4.2.17 *Let $B = B_\tau \otimes R$ be an irreducible indecomposable Murley group. Then the following hold:*

i) *A rank-1 group A belongs to ${}^*B$ if and only if $A \otimes B_0 \cong \mathbb{Q}$ or $tp(A) \leq \tau$.*

ii) *A countable torsion-free group A belongs to ${}^*B$ if and only if there exists an ascending chain of B-cobalanced subgroups $\{A_n\}_{n<\omega}$ of A such that $A_0 = 0$ and A_{n+1}/A_n is a rank-1 group, whose type is less than or equal to τ, or $(A_{n+1}/A_n) \otimes B_0 = \mathbb{Q}$.*

Proof: i) Let $A \leq \mathbb{Q}$ with $\mathrm{Ext}(A, B)$ torsion-free. Without loss of generality, we may assume that A is a B_0-module. If $\mathrm{Hom}(A, B) = 0$, then $A \cong \mathbb{Q}$ by Proposition 1.7.9. On the other hand, if there exists a non-zero homomorphism $\alpha : A \to B$, then α is one-to-one and $tp(A) \leq tp(\alpha(A)_*) = \tau$.

Conversely, assume that A is a rank-1 B_0-module with $tp(A) \leq \tau$. Without loss of generality, $B_0 \leq A$. Since A is a B_0-module and B is homogeneous, we have $A = pA$ if and only if $B = pB$. If U is a subgroup of $\mathbb{Q}$ of type τ which contains A, then U/B_0 is a reduced torsion group, and the same holds for A/B_0. Since $tp(A) \leq \tau$, we obtain $\mathrm{Hom}(A, B) \neq 0$. Consider the induced sequence

$$0 = \mathrm{Hom}(U/A, B) \to \mathrm{Hom}(U, B) \to \mathrm{Hom}(A, B)$$

$$\xrightarrow{\delta} \mathrm{Ext}(U/A, B) \to \mathrm{Ext}(U, B) \to \mathrm{Ext}(A, B) \to 0.$$

Since U is isomorphic to a pure subgroup of B, $\mathrm{Ext}(U, B)$ is torsion-free. Therefore $\mathrm{Ext}(U/A, B)/\mathrm{Im}(\delta)$ is torsion-free. However, U/A is a direct sum of cyclic groups, so that $\mathrm{Ext}(U/A, B) \cong \prod_{p \in \mathbb{P}} B/p^{n_p} B$ for suitable $n_p < \omega$. Therefore $t(\mathrm{Ext}(U/A, B)) \leq$

Im(δ) and $\mathrm{Ext}(U/G, B)/\mathrm{Im}(\delta)$ is divisible. Consequently, $\mathrm{Ext}(A, B)$ is isomorphic to a direct summand of the torsion-free group $\mathrm{Ext}(U, B)$ and so it is itself torsion-free.

ii) Suppose that $\mathrm{Ext}(A, B)$ is torsion-free and write A as the union of a chain of pure subgroups A_n with $r_0(A_n) = n$. By Theorem 2.2.2, each A_n is B-cobalanced in A. Hence $A_{n+1}/A_n \in {}^*\mathcal{B}$ and so it has the desired form by Part i).

The converse holds by Theorem 3.1.4. $\square$

Corollary 4.2.18 *Let B be a p-local irreducible indecomposable Murley group. Then $\mathrm{Ext}(A, B)$ is torsion-free for all finite rank Butler groups A.*

Proof: Since B is p-local, $B_0 = \mathbb{Q}_p$. If A is a Butler group, then $A \otimes \mathbb{Q}_p$ is completely decomposable and so $\mathrm{Ext}(A, B)$ is torsion-free. $\square$

We next show that, even in the case of a Murley group B of rank 2, there may exist groups A with $K_B(A) = 0$ but $\mathrm{Ext}(A, B)$ not torsion-free.

Theorem 4.2.19 *Let B be torsion-free such that $E = \mathrm{End}(B)$ is hereditary and $\mathrm{Ext}(B, B)$ is torsion-free. Then the following are equivalent:*

i) *$A \in {}^*\mathcal{B}$ for every group A with $K_B(A) = 0$;*

ii) *Whenever $K_B(A) = 0$ then $S_B(A)$ is a direct summand of A;*

iii) *If $0 \to U \to A \to H \to 0$ is a short exact sequence with $U \leq B$ and $\mathrm{Ext}(H, B)$ torsion-free, then $\mathrm{Ext}(A, B)$ is torsion-free provided $K_B(A) = 0$.*

Proof: i) $\Rightarrow$ ii) Since $\mathrm{Ext}(B, B)$ is torsion-free, B is a finitely faithful S-group and $S_B(A)$ is a pure subgroup of A. Thus $S_B(A)$ is a quasi-summand of $\bigoplus_n B$, whenever $A \leq \bigoplus_n B$, say
$$k \cdot \bigoplus_n B \leq S_B(A) \oplus U \leq \bigoplus_n B$$
for some non-zero integer k. Then $kA \leq S_B(A) \oplus (U \cap A) \leq A$ and hence
$$A/S_B(A) \cong k \cdot (G/S_B(A)) = (kA + S_B(A))/S_B(A) \leq U \cap A$$
has zero B-radical. Moreover, $S_B(A)$ is B-projective as a B-generated subgroup of $\bigoplus_n B$ since E is hereditary. Thus the group $\mathrm{Ext}(A/S_B(A), S_B(A))$ is isomorphic to a direct summand of $\bigoplus_l \mathrm{Ext}(A/S_B(A), B)$ for some $l \in \mathbb{N}$ and the latter is torsion-free by i). However, the short exact sequence
$$0 \to S_B(A) \to A \to A/S_B(A) \to 0$$

represents a torsion element of $\text{Ext}(A/S_B(A), S_B(A))$, which is a contradiction unless the sequence represents the zero element in $\text{Ext}(A/S_B(A), S_B(A))$. But this means nothing else but $S_B(A)$ is a direct summand of A.

ii) $\Rightarrow$ i) We consider a short exact sequence

$$0 \to B \xrightarrow{\alpha} H \to A \to 0,$$

which is quasi-splitting. Then $K_B(H) = 0$ and $S_B(H)$ is a direct summand of H by ii). Therefore $\alpha(B)$ is a pure subgroup of $S_B(H)$, which is itself B-projective since E is hereditary. We obtain the induced exact sequence

$$0 \to \text{Hom}(B, A) \xrightarrow{\alpha_*} \text{Hom}(B, H) \to M \to 0,$$

in which M is a finitely generated submodule of $H_B(A)$, because $\text{Hom}(B, H) \cong \text{Hom}(B, S_B(H))$ is projective. Thus M is projective and the sequence splits. We get the commutative diagram

$$\begin{array}{ccc}
B \otimes \text{Hom}(B, B) & \to & B \otimes \text{Hom}(B, H) \\
\downarrow \theta_B & & \downarrow \theta_H \\
0 \to \quad B & \xrightarrow{\alpha} & S_B(H)
\end{array}$$

where the maps θ_B and θ_H are isomorphisms. Hence $\text{Im}(\alpha)$ is a direct summand of $S_B(H)$ and thus the short exact sequence

$$0 \to B \xrightarrow{\alpha} H \to A \to 0$$

is already split-exact, which implies that $\text{Ext}(A, B)$ is torsion-free.

Since i) $\Rightarrow$ iii) is trivial, it remains to show the converse. If $K_B(A) = 0$ then $A \leq \bigoplus_n B$. By iii) the group $A \cap B$ satisfies $\text{Ext}(A \cap B, B)$ is torsion-free. Moreover, $A/(A \cap B) \cong (A+B)/A \leq \bigoplus_{n-1} B$. By induction hypothesis, also $\text{Ext}(A/(A \cap B), B)$ is torsion-free. Using iii) again, we deduce $\text{Ext}(A, B)$ is also torsion-free. $\square$

Our next example shows, that the condition on $S_B(A)$ being a direct summand cannot be dropped.

Example 4.2.20 *Let p, q and r be three distinct primes and consider the groups $B = \mathbb{Q}_{p,r}$, $C = \mathbb{Q}_{p,q}$ and $D = \mathbb{Q}_r$. Then $A = C \oplus D$ is obviously a Murley group. By [Ar] the group $B \oplus C$ contains an indecomposable subgroup G such that $|(B \oplus C)/G| = p$. Since $B \oplus C \leq A$, we have $K_A(G) = 0$. Moreover, $S_A(G) \leq S_A(B \oplus C) = C$ since $\text{Hom}(D, B) = \text{Hom}(D, C) = 0$. Thus $r_0(S_A(G)) = 1$, but $S_A(G)$ is not a direct summand of G. By the last theorem, A does not have the extension property.*

We finish with a short remark on the general torsion-free case:

Proposition 4.2.21 *The following holds in the lattice* $(\mathfrak{Tf}, \mathfrak{Tf})$-$\mathcal{TF}$:

i) $\mathbf{1} = ((\mathfrak{F} \oplus \mathfrak{D}) \cap \mathfrak{Tf}, \mathfrak{Tf})$

ii) $\mathbf{0} = (\mathfrak{Tf}, \mathfrak{D} \cap \mathfrak{Tf})$

Proof: Clear. □

Finally, we calculate the $(\mathfrak{Tf}, \mathfrak{Tf})$-torsion-free pairs cogenerated and generated by $\mathbb{Z}$, as an example. Here we clearly have $\mathbb{Z}^* = \mathfrak{Tf}$ and hence the $(\mathfrak{Tf}, \mathfrak{Tf})$-torsion-free pairs cogenerated by $\mathbb{Z}$ is the maximal one, $((\mathfrak{F} \oplus \mathfrak{D}) \cap \mathfrak{Tf}, \mathfrak{Tf})$.

Determining $^*\mathbb{Z}$ we refer the reader on the concept of coseparable groups. Here we directly see that $^*\mathbb{Z} = (\mathfrak{Cos} \oplus \mathfrak{D}) \cap \mathfrak{Tf}$.

Bibliography

[Al1] **U. Albrecht**, *Endomorphism rings and A-projective torsion-free abelian groups*, Lect. Notes Math., Vol. **1006**, Springer-Verlag Berlin-Heidelberg-New York, 1983, 209-227.

[Al2] **U. Albrecht**, *Abelian groups, A, such that the category of A-solvable groups is preabelian*, Contemp. Math. **87**, 1989, 117-131.

[Al3] **U. Albrecht**, *Locally A-projective abelian groups and generalizations*, Pacific J. Math. **141**, 1990, 209-228.

[Al4] **U. Albrecht**, *Two-sided essential submodules of $\mathbb{Q}^r(R)$*, Houston J. Math. **33**, 2007, 103-123.

[Al5] **U. Albrecht**, *Endomorphism rings of faithfully flat abelian groups*, Resultate Math. **17**, 1990, 179-201.

[AlGoe] **U. Albrecht, P. Goeters**, *Butler theory over Murley groups*, J. Algebra **200**, 1998, 118-133.

[AlGoe2] **U. Albrecht, P. Goeters**, *Strong S-groups*, Colloquium Math. **80**, 1999, 97-105.

[Ar] **D.M. Arnold**, *Finite rank torsion-free abelian groups and rings*, Lect. Notes Math. Vol. **931**, Springer-Verlag Berlin-Heidelberg-New York, 1982.

[Ar2] **D.M. Arnold**, *Endomorphism rings and subgroups of finite rank torsion-free abelian groups*, Rocky Mountain J. Math. **12(2)**, 1982, 241-256.

[ArMu] **D.M. Arnold, C.E. Murley**, *Abelian groups, A, such that $Hom(A,-)$ preserves direct sums of copies of A*, Pacific J. Math. **56(1)**, 1975, 7-20.

[B] **R. Baer**, *Erweiterungen von Gruppen und ihren Isomorphismen*, Math. Z. **38**, 1933, 375-416.

[BaSa] **S. Bazzoni, L. Salce**, *An independence result on cotorsion theories over valuation domains*, J. Algebra **243**, 2001, 294-320.

[EkHu] **P.C. Eklof, M. Huber**, *On the rank of Ext*, Math. Z. **174**, 1980, 159-185.

[EkMe] **P.C. Eklof, A.H. Mekler**, *Almost free Modules*, Vol. **46**, North Holland Mathematical Library, 1990.

[FaGoe] **T.G. Faticoni, H.P. Goeters**, *On torsion-free* Ext, Comm. Algebra **16(9)**, 1988, 1853-1876.

[Frie] **S. Friedenberg**, *Eine Charakterisierung von* $\text{Ext}(A, H)$ *für abzählbare abelsche Gruppen A und H*, Diplom-Thesis, University of Duisburg-Essen, 2008.

[Fu1] **L. Fuchs**, *Infinite Abelian Groups*, Vol. **1**, Academic Press, New York and London, 1970.

[Fu2] **L. Fuchs**, *Infinite Abelian Groups*, Vol. **2**, Academic Press, New York and London, 1973.

[G] **K. Goodearl**, *Non-Singular Rings and Modules*, Marcel Dekker, 1976.

[GöSh] **R. Göbel, S. Shelah**, *Cotorsion theories and splitters*, Trans. AMS **352**, 2000, 5357-5379.

[GöShWa] **R. Göbel, S. Shelah, S. Wallutis**, *On the lattice of cotorsion theories*, J. Algebra **238**, 2001, 292-313.

[Goe] **H.P. Goeters**, *When is* $\text{Ext}(A, B)$ *torsion-free? and related problems*, Comm. Algebra **16(8)**, 1988, 1605-1619.

[Goe1] **H.P. Goeters**, *Generating cotorsion theories And injective classes*, Acta Math. Hung. **51(1-2)**, 1988, 99-107.

[Hau] **J. Hausen**, *Automorphismen gesättigte Klassen abzählbarer abelscher Gruppen*, Studies on abelian Groups, Springer, Berlin, 1968, 147-181.

[HuWa] **M. Huber, R.B. Warfield, Jr.**, *On the torsion subgroup of* $\text{Ext}(A, B)$, Archiv Math. **32**, 1979, 5-9.

[HuWa2] **M. Huber, R.B. Warfield, Jr.** *Homomorphisms between cartesian powers of abelian groups*, Abelian Group Theory, Proceedings Oberwolfach, 1981, 202-227.

[Hül] **N. Hülsmann**, *Cotorsion pairs for Bext and a generalization of Whitehead´s problem*, Ph.D.-Thesis, Essen, 2006.

[Ma] **A. Mader**, *Almost Completely Decomposable Groups*, Algebra, Logic and Applications **2**, Gordon and Breach, Amsterdam, 2000.

[Ma1] **A. Mader**, *The group of extensions of a torsion group by a torsion-free group*, Archiv Math. **20**, 1969, 126-131.

[MeSh] **A. Mekler, S. Shelah**, *Every coseparable group may be free*, Israel J. Math. **81**, 1993, 161-178.

[Ro] **J. Rotman**, *On a problem of Baer and a problem of Whitehead in abelian groups*, Acta Math. Acad. Sci. Hung. **12**, 2000, 245-254.

[Sa] **L. Salce**, *Cotorsion theories for abelian groups*, Symp. Math. **23**, 1977, 11-32.

[Sch] **P. Schultz**, *Self-splitting groups*, Preprint series of the University of Western Australia at Perth, 1980.

[Sh1] **S. Shelah**, *Whitehead groups may not be free even assuming CH, I*, Isr. J. Math. **28**, 1977, 193-204.

[Sh2] **S. Shelah**, *Whitehead groups may not be free even assuming CH, II*, Isr. J. Math. **35**, 1980, 257-285.

[ShStr] **S. Shelah, L. Strüngmann**, *The p-rank of $\mathrm{Ext}_{\mathbb{Z}}(G,\mathbb{Z})$ in certain models of ZFC*, Algebra and Logic **46**, 2007, 369-397.

[St] **B. Stenström**, *Rings of Quotients*, Lect. Notes Math. **217**, Springer Verlag, Berlin, Heidelberg, New York, 1975.

[Str] **L. Strüngmann**, *(V,W)-cotorsion pairs*, Archiv Math. **86**, 2006, 193-204.

[Wa] **R.B. Warfield, Jr.**, *Extensions of torsion-free abelian groups of finite rank*, Archiv Math. **23**, 1972, 145-150.

[Wa1] **R.B. Warfield, Jr.**, *Homomorphisms and duality for torsion-free groups*, Math. Z. **107**, 1968, 189-200.

Die VDM Verlagsservicegesellschaft sucht für wissenschaftliche Verlage abgeschlossene und herausragende

Dissertationen, Habilitationen, Diplomarbeiten, Master Theses, Magisterarbeiten usw.

für die kostenlose Publikation als Fachbuch.

Sie verfügen über eine Arbeit, die hohen inhaltlichen und formalen Ansprüchen genügt, und haben Interesse an einer honorarvergüteten Publikation?

Dann senden Sie bitte erste Informationen über sich und Ihre Arbeit per Email an *info@vdm-vsg.de*.

Sie erhalten kurzfristig unser Feedback!

VDM Verlagsservicegesellschaft mbH
Dudweiler Landstr. 99   Telefon +49 681 3720 174
D - 66123 Saarbrücken   Fax   +49 681 3720 1749
www.vdm-vsg.de

Die VDM Verlagsservicegesellschaft mbH vertritt

Printed by Books on Demand GmbH, Norderstedt / Germany